"Fusing the scholar's pen and the surgeon's scalpel, *The Body* exposes the intellectual innards of the most taken-for-granted unit of social life. With verve and originality, the text takes seemingly discrete objects—genes, cells, organs, and more—and carefully situates them within their layered biopolitical contexts. This is essential reading for all those who want conceptual tools to engage, challenge, and reimagine the nature/nurture binary."

Ruha Benjamin, Princeton University, Author of
People's Science: Bodies and Rights on the Stem Cell Frontier

"This highly original and exciting textbook significantly updates the sociology of the body while also offering a new style of engagement with readers from many disciplines. A great strength is its accessibility, making available complex theoretical and conceptual analyses of bodies and embodiment through vivid case studies that bring this subject alive and demonstrate its relevance to every aspect of contemporary society."

Professor Sarah Franklin, Wellcome Trust Senior Investigator and Director Reproductive Sociology Research Group (ReproSoc), University of Cambridge

"*The Body: Social and Cultural Dissections* is a work of subtle genius that teaches us to rethink the very things we think we know the most about: our bodies and ourselves. Brilliantly organized and thoughtfully written, the book concisely anatomizes how bodies are shaped, deployed, and politicized into symbols of deeper meaning. This is the rare book that synthesizes theory, history, and volumes of information into a single, compelling narrative. *The Body* is required reading for students, teachers, practitioners, and anyone else who wants to learn more about the valences and implications of human existence."

Jonathan Metzl, Director, Center for Medicine, Health, and Society, Frederick B. Rentschler II Professor of Sociology, Vanderbilt University

"Moore and Casper have written a fantastic book. Organized around body parts and functions, the book brilliantly turns anatomy on its head, showing how the body is not stable and natural, but malleable, transformable, and always enmeshed in power relations. It is a superb contribution to the interdisciplinary literature on the body. Absolutely a must-read for both scholars and students."

Jonathan Xavier Inda, University of Illinois, Urbana-Champaign,
Author of Racial Prescriptions: Pharmaceuticals, Difference, and the Politics of Life

"Moore and Casper have successfully taken on the project of fleshing out anatomy to reveal the social and cultural meanings that complicate and partly construct our biological understandings. *The Body* is firmly situated in critical body studies, read mainly through a sociological perspective but always aware of, and responsive to, the interdisciplinarity of the field. It is a valuable and engaging guide for students and teachers alike."

Margrit Shildrick, Professor of Gender and Knowledge Production,
Linköping University

The Body

This college-level handbook offers a comprehensive and accessible overview of sociological and cultural perspectives on the human body. Organized along the lines of a standard anatomical textbook delineated by body parts and processes, this volume subverts the expected content in favor of providing tools for social and cultural analysis.

Students will learn about the human body in its social, cultural, and political contexts, with emphasis on multiple, contested meanings of the body, body parts, and systems. Case studies, examples, and discussion questions are both US-based and international. Advancing critical body studies, the book explicitly discusses bodies in relation to race, class, gender, sexuality, ability, age, health, geography, and citizenship status. The framing is sociological rather than biomedical, attentive to cultural meanings, institutional practices, politics, and social problems. The authors use commonly understood anatomical frames to discuss social, cultural, political, and ethical issues concerning embodiment.

Lisa Jean Moore, a medical sociologist, is Professor of Sociology and Gender Studies at Purchase College, State University of New York. Her scholarship is located at the intersections of sociology of health and medicine, science and technology studies, feminist studies, animal studies, and critical body studies. She is the author of *Sperm Counts: Overcome by Man's Most Precious Fluid* and the co-author of *Gendered Bodies: Feminist Perspectives*, *Missing Bodies: The Politics of Visibility*, and *Buzz: Urban Beekeeping and the Power of the Bee*. Additionally she has co-edited *The Body Reader* and is a founding co-editor of a successful book series at NYU Press entitled *Biopolitics: Medicine, Technoscience, and Health in the 21st Century*.

Monica J. Casper is Professor and Head of Gender and Women's Studies at the University of Arizona. A sociologist, her scholarly and teaching interests include gender, race, bodies, health, sexuality, disability, and trauma. She has published several books, including the award-winning *The Making of the Unborn Patient: A Social Anatomy of Fetal Surgery*. Her current research focuses on race and the biopolitics of infant mortality in the U.S. She is a founding co-editor of the NYU Press book series *Biopolitics: Medicine, Technoscience, and Health in the 21st Century*, as well as a managing editor of *The Feminist Wire*.

The Body

Social and Cultural Dissections

Lisa Jean Moore and Monica J. Casper

Routledge
Taylor & Francis Group
LONDON AND NEW YORK

First published 2015
by Routledge
2 Park Square, Milton Park, Abingdon, Oxon OX14 4RN

and by Routledge
711 Third Avenue, New York, NY 10017

Routledge is an imprint of the Taylor & Francis Group, an informa business

British Library Cataloguing-in-Publication Data
A catalogue record for this book is available from the British Library

Library of Congress Cataloging-in-Publication Data
Moore, Lisa Jean, 1967–
The body : social and cultural dissections / by Lisa Jean Moore and Monica J. Casper.
pages cm
1. Human body—Social aspects. 2. Human body (Philosophy) 3. Human anatomy—Social aspects. I. Casper, Monica J., 1966– II. Title.
HM636.M66 2014
128'.6—dc23 2014014901

ISBN: 978-0-415-82195-7 (hbk)
ISBN: 978-0-415-82196-4 (pbk)
ISBN: 978-0-203-55683-2 (ebk)

Typeset in Frutiger
by Keystroke, Station Road, Codsall, Wolverhampton

Contents

List of figures

Acknowledgments

We are indebted to the international anonymous reviewers who enhanced and improved this book with their keen vision. Joe Osmundson provided helpful scientific expertise. Editorial assistant Alyson Claffey's discerning eye and intelligence has also been greatly appreciated. And we are grateful to our families for granting us time to write.

chapter 1

Introduction

Snapshots of the 21st century

Is the image shown on page 2 a cyborg soldier from the movie *Transformers*, or perhaps the Tony Stark character from *Iron Man*? Or is it real life?

The U.S. military has been testing the Universal Human Load Carrier (HULC), an exoskeleton designed to enhance physical capability, including strength and endurance. Such enhancement could certainly yield a more resilient soldier.

Funded by Lockheed Martin with testing supported by the military, the technology (like so many originally developed through military–industry collaborations) has applications far beyond the military.

For example, Figure 1.2 depicts a child born with arthrogryposis multiplex congenita (AMC), a genetic condition that causes muscles and joints to be stiff and immobile, fitted with an exoskeleton. With this "enhancement," she is able to move her limbs, hug her mother, carry objects, and otherwise function with increased mobility.

As sociologists, we are fascinated by these kinds of technologies. Developed in one context for a specific set of purposes, they are subsequently reconfigured for other kinds of bodies and activities. These images, which resemble characters in science fiction films, now reflect varied human realities. One exemplifies the reach of the U.S. military and a multibillion-dollar defense budget; the other emphasizes the domestic sphere of family life, shared embodied intimacy, and the quest for a "normal" childhood.

These examples of extreme uses of bodily technology illustrate that human anatomy is changeable. That is to say, bodies can be transformed and manipulated. Unlike the staid, two-dimensional drawings of human bodies found in most medical textbooks, the images here point to dynamic

FIGURE 1.1 A model "Soldier" tries out the exoskeleton
Photo courtesy of Lockheed Martin

FIGURE 1.2 Little girl hugging mother
Photo courtesy of Stratasys

anatomies that can be upgraded, fixed, or reordered through human ingenuity and technical know-how. But while these are "high-tech" examples, humans also change their anatomies through seemingly "low-tech" practices, such as application of sunscreen and hair color, exercise, eating, and drinking water.

In opening this book, we have deliberately selected spectacular examples from contemporary life in the Global North to highlight certain points about the malleability of bodies. And yet, the impetus for the book is to explore and explain the many ways in which human bodies are culturally shaped, variously deployed, relentlessly political and politicized, sources of potentials and limitations, simultaneously material and symbolic, subject to interventions and transformations, finite and flexible, powerful and vulnerable, and endlessly open to varied, contested representations, and interpretations.

Bodies are anything but simple.

Structure of the book

This book is organized around body parts and functions: anatomy. This may seem like an unusual strategy for a book based in the social sciences and humanities. After all, anatomy is the scientific study of the structure of organisms and is rooted in ancient civilizations. Anatomical ideas appeared as early as 1600 BCE, and expanded through development of the printing press. Anatomy and its representations offered new understandings of human material existence.

Yet, from the beginning, anatomy was firmly grounded in actual dissections and examinations of living and dead organisms, including human bodies. The focus was on the body itself, and not on the person, and the goal was to develop physiological and biological knowledge about bodies and their functions. Anatomical understandings have underpinned everything from biology to medicine to forensic pathology. Anatomy has been foundational for many scientific and popular representations of the body. For example, we speak of "heads of state" in politics and the "long arm" of the law.

We argue, however, that these anatomical renderings are limited in their reliance on stable, static, allopathic, normative, derivative, and standard notions of bodies. Anatomical examinations of bodies are, for the most part, asocial; presented as if bodies are not engaged in cultural, political, and kinetic practices every second of every day. Here, we take up the practice of dissection but we do so sociologically to repurpose anatomy. We use the conventions of standard anatomy to identify body parts and processes, but we examine these through a lens that reveals their social and cultural meanings. We enflesh and animate the flattened human bodies of anatomy textbooks by situating them in social, cultural, and geopolitical contexts.

In this book, dissection does not mean surgically slicing into flesh and organs, but rather *conceptually* and *theoretically* opening up human bodies. Our methods—sociological and cultural analysis—allow for sharp, precise, and systematic investigation of layers of fascia, meaning, social organization, and transnational flows. Just as the body is comprised of multiple layers—molecular, fluid, bony, fleshy, interactions between bodies—our analysis addresses the stratified and interconnected layers of society. We are deeply attentive to the body in a social context, including persistent inequalities of gender, race, sexuality, age, ability, geography, and other markers of human status.

Building on the work of scholars such as Donna Haraway, Londa Schiebinger, Adele Clarke, and others who have critically examined disciplinary foundations of biology, we begin with the premise that human anatomy—biology—provides only a limited perspective on the body. Sociologists and others study the body because it has been a key site for power operations, with bodies classified and targeted for both damaging and liberatory ends. We show in this book that social scientists and humanists have a claim on the body; that bodies are not the sole province of the natural sciences. We consider bodies from multiple angles. Rather than focusing exclusively on function, we consider how bodies produce real effects in the world, how bodies are impacted by structural and material conditions, and how bodies are represented in multiple, diverse contexts.

This introduction provides an overview of what we are calling *critical body studies*. For some 30 years, various fields (e.g., anthropology, sociology, philosophy, literary studies, and history) have examined the body as a component of social relations. Yet only very recently has critical body studies come to be recognized and institutionalized as an area of study in its own right through conferences, journals, and centers. This textbook deliberately and self-consciously marks a moment in the formation of critical body studies as a distinct intellectual field with relevance across a number of other fields, including our home discipline of sociology.

Our introduction proceeds with a moment about being embodied scholars (and learners), and then moves through a range of theoretical approaches to the body. We then offer a brief overview of critical body studies, including its major ideas and theorists, before moving on to a discussion of several interdisciplinary fields related to and, in some cases dependent on, critical body studies. We conclude the introduction with a set of learning objectives for the book, definitions of core concepts, and an owner's manual.

Our reflexive moment

While working in the field of critical body studies as scholars *and* human beings, we are relentlessly aware of our own bodies, as well as those of our intimates. Collaborating on this book, as in our earlier collaborations (Casper and Moore 2009), has ignited a fresh awareness of our own body projects. We are middle-aged, highly educated, white women living in the U.S., mothers to five girls with their own body anxieties, and mentors to numerous students who engage in diverse body projects including cosmetic surgery, gender-affirming surgeries, body modification practices, and dieting and starvation.

We have bodies, and we are alternately reluctantly and willfully complicit in body projects and anxieties. For example, we use a variety of facial cleansers and creams to keep our skin looking youthful (especially Monica, who lives in the desert), we dye our hair, we have tattoos, we shave, we have given birth (three vaginal births, two cesarean sections), we have had cancer, we have undergone various surgeries for broken ankles and ovarian cysts and other mishaps, we have buried family members, we have experienced miscarriage and abortion and prenatal testing, we have worn false eyelashes and make-up, and we have engaged in numerous physical activities ranging from sex to sports, sleeping to eating.

And because we are also sociologists, these embodied activities ignite critical reflection and dialogue about broader institutional and interactional frameworks within which our everyday lives unfold. In other words, our own embodied experiences in the world inform and shape our research expertise.

For example, Lisa Jean's reproductive experiences led her to a decade-long study of human semen, sperm banking, and somatic representations of masculinity—she even became the Board President of a California nonprofit sperm bank for six years. Inspired by the feminist self-help health movement, Lisa Jean spent several years exploring 20th century anatomical renderings of female and male genitalia. She has also published scholarly research about gender and birth certificate policies, breastfeeding, sex work and safer sex, and urban homesteading in New York City.

An expert on women's health, Monica has written about cervical cancer, the Pap smear, the HPV vaccine, abortion, contraception, pregnancy, reproductive politics, breast cancer, genetic testing, fetal tissue research, infant mortality, health disparities, biomedical technologies, environmental health, disability, and traumatic brain injury. She won awards for her 1998 book *The Making of the Unborn Patient: A Social Anatomy of Fetal Surgery*, and is a frequent commentator in public media on reproductive health and justice issues. She often writes about her own reproductive experiences, including abortion, two cesarean sections, and mothering young daughters. With Lisa Jean, Monica has written extensively about how human bodies are rendered invisible, visible, or hypervisible in public discourse.

The experience of being an embodied learner is necessarily shared by everyone reading this book. We all have bodies. But because no two bodies are alike, our embodied experiences are varied, and they are specific to the reader's age, race, ethnicity, gender, sexuality, class status, ability, family history, culture, and geographic location.

Challenging the "natural" body

When asked about inequities between men and women with respect to economic and political power, our students often claim exasperatedly: "Women's and men's bodies are just *different*!" These presumed differences are cited as evidence to explain why our culture (and that of many others) is organized in certain ways. For example, almost universally women are responsible for care of children (whether their own or those of others) and the maintenance of the domestic or household domain, as it is presumably in women's "nature" to be domestic. But this so-called natural arrangement is historical, ideological, and also unsustainable, as women have entered the paid labor force in greater numbers beginning in the latter half of the 20th century. The shifts have produced considerable tensions (Hochschild 2012 [1989]).

Importantly, scholars of the body have worked to reveal how these "self-evident" differences are actually culturally produced. Certain bodies survive and thrive depending on economic resources and social power, while others become ill or even die. For example, throughout the world, men's bodies are at risk of military, athletic, and industrial exploitation, and, for disadvantaged men, imprisonment and premature, often violent, death. Institutions dominated by men, namely medicine and religion, disproportionately control women's bodies; as struggles to attain comprehensive access to birth control, abortion, and safe maternity care demonstrate, reproductive bodies are particularly targeted.

Through discourses and practices of science and medicine, including human anatomy, these differences are recast as natural, physical, universal, trans-historical, and permanent. It is commonly understood in tautological fashion that men's and women's bodies are different because they were born that way. Furthermore, a vast majority of anatomical illustration has been based on male bodies (specifically male cadavers), often without acknowledgment of the gender of the model – rather, the male body is taken as universal. Feminist activism and scholarship has increased awareness of how bodies are *gendered* by making visible the cultural and social dynamics that produce difference and dominance out of the flesh of bodies (Van den Wijngaard 1997).

In effect, many feminist scholars have challenged the notion of the "natural" body. In general, naturalistic or biological approaches to the body hold that humans—including identities—are constrained and/or enabled by their birth-given characteristics (e.g., sex, skin color, height, genitals); this is a form of *biological determinism*. Correspondingly, social relationships, institutions, and the ideologies that impact human life are founded upon particular assumptions about the biological body. What the body looks like is thought to reveal who someone is, and thus forms the basis for how people are identified socially and legally.

Naturalistic approaches to the body have produced a highly charged field of work, particularly revolving around the sociology of gender and claims about the physiological basis of women's inequality. Some of the earliest and most controversial work in this vein emanated out of sociobiology in the 1970s, particularly the work of Harvard sociobiologist Edward O. Wilson, who received much praise and publicity for his theories of genetic evolution and social behavior.

According to Wilson (1975, 1979), human behavior is explained by and encoded within the gene. Wilson attempted (many argue unsuccessfully) to link genetic structures in animals to those in humans to establish a biological basis for human behavior. Wilson maintains that women's social subordination was "natural" because "women as a group were less assertive and physically aggressive" owing to their genetic make-up (Wilson 1978: 128).

Not coincidentally, sociobiology developed simultaneously with the rise of the women's movement in the Global North, particularly radical feminism. The field of sociobiology quickly became a useful way for social conservatives to undermine the expansion of feminist discourses and calls for gender inequality. Sociobiological beliefs continue today, and "women's place" remains a contested topic, with conservatives typically linking women's subordinate status to biological functions such as reproduction.

Ideas frequently circulate between biology and sociology. Like Wilson, mid-century American sociologist Talcott Parsons (1964) also attributed female inequality to biology. In his structural functionalist framing, woman is an inherently natural creature (best suited for reproductive work), while man, the more cultured being, belongs in the public sphere (the world of production).

Parsons further maintained that female discrimination in the workplace is functional to society because occupational equality is "incompatible with any positive solidarity in the family" (Parsons 1940: 852). For Wilson the biologist and Parsons the sociologist, the "woman question" is conveniently answered in one word—biology. Anatomy is destiny.

Similar to functional and neo-social Darwinist theories, some early feminist theory also prioritized corporeal analyses in explanations of patriarchy and women's position in society. Female biology, for these theorists, accounts for why women behave in certain ways or have particular proclivities toward social engagement. For example, see Rich (1976) and Chodorow (1978) for approaches that provide naturalistic explanation for social patterns regarding reproduction and reproductive work—i.e., mothering.

Radical feminist Shulamith Firestone's *The Dialectic of Sex* (1970) serves as a prime example of feminist biological reductionism. While Firestone recognized that social institutions assist in maintaining patriarchy, ultimately the foundations of male dominance reside in women's reproductive capabilities. She famously suggested that if women could be free of reproduction, they could be free.

Both functionalist and feminist arguments emphasizing biology are highly problematic. In these accounts, corporeality subsumes or erases sociality, and the social construction of gender—that is, the ways in which gender is inextricably social and created through our relations—is typically neglected or absent altogether.

A notable and useful work on the body and biology within the sex/gender debate is Thomas Laqueur's *Making Sex* (1990), a historical account of the medical, political, and cultural construction of sex from ancient Greece to the Enlightenment. Unlike the above arguments that understand gender differences as a result of biology (sex), Laqueur traces how medical knowledge and common understandings of sex and sexuality were based on *cultural* discourses rather than biology.

To simplify, he illustrates how our biology was, and continues to be, culturally determined. According to Laqueur, the corporeal body (visible flesh and blood) cannot be regarded as the "real" foundation for cultural claims because cultural meanings already underpinned the biology. This assertion is evidenced in the rather extraordinary one-sex model of the body that held sway as a biological given until the end of the 17th century. Galen's "male model or unborn penis model" stated the "obvious," that women had an unborn penis inside their bodies, thus proving biologically women's lesser perfection.

Laqueur's work reminds us that our duty as social scientists and historians is to understand how what we understand as biologically "real" is a material expression of other, more pervasive, culturally constructed meanings. It is through the institutional and persuasive power of science that these meanings become elevated as truth.

Scholars have also explored how current biological ideas are culturally constructed and exist along a historical continuum. Historian and sociologist Barbara Duden's *Disembodying Women* (1993) examines the historical processes by which pregnancy has been transformed from a personal experience—between a woman and her fetus/baby—to an impersonal and even public concern via technological advancements in medicine and shifts in discourses.

As Casper (1998) argued with respect to fetal surgery almost two decades ago, the emergence of the fetus as a distinct patient and person has led to a corresponding decrease in the autonomy and rights of pregnant women, both in medicine and the political sphere. This is significant, for as Moira Gatens (1996) notes, women are already often forced to "elide" or suppress their own "corporeal specificity" to participate in liberal democracies.

The self-help women's health movement especially contested predominant biomedical approaches to human bodies (Ruzek 1978; Lewin and Olesen 1985). Using both their power as consumers and their increasing access to academia, many women rebelled against the hegemonic medical establishment's strategies of medicalization and mystification of female bodily functions. These challenges to "thinking as usual" within medical settings encouraged many women to wage feminist critiques against the standardization of male bodies as the model for individualism and better health.

Emblematic of this change was publication of *Our Bodies, Ourselves* and *A New View of a Woman's Body*, which emerged as owner's manuals to women's bodies, written by women for women. Historian Kathy Davis (2007) describes *Our Bodies, Ourselves*—sometimes called the "women's health bible"—as an epistemological project that "can help 'ground' feminist theoretical discussions more firmly in women's sentient and situated knowledge. More specifically, it provides avenues for the development of feminist body theories that are *less* disembodied, *less* disembedded, and *more* relevant to women's health concerns" (140, emphasis in original).

Key to various women's health perspectives has been a challenge to biomedical framings of women's bodies (Clarke and Olesen 1998). Rather than seeing the body and its processes as pathological, women's health scholars have insisted on revealing the social, political, and cultural operations that shape notions of health and illness, both normal and abnormal.

For example, contemporary scholarship on menstruation shows how "periods" have become pathologized, with women prescribed birth control pills to regulate the monthly cycle and "feminine hygiene" products branded using medicalized discourses (Stein and Kim 2009). Women's health scholarship identifies the stakes of defining women's bodies in particular ways and advocates for structural, cultural frameworks to supplement biological knowledge.

Like critical scholarship in women's health research, feminist science and technology studies (STS) also challenges dominant biological and biomedical definitions of health and bodies. Biologist and historian Donna Haraway (1991) is perhaps the most renowned feminist STS scholar. Her work empirically exposes layers of meaning and politics underlying sex/gender distinctions and the role of sex/gender in shaping who and what counts as human. Her ideas, in particular cyborg theory, offer us new ways to interrupt dichotomous constructions of nature and culture, human and nonhuman, and organic and inorganic matter. This work, while taking account of materiality and the obduracy of planetary forms and processes, also nonetheless shows how what we know as "truth" and "nature" is historically, socially, and technologically shaped.

Inspired by Foucault and political theory, Judith Butler (1993) represents a poststructural account of bodies. Butler proposes a theory of *materialization* to counter the absence of the physicality of bodies in much social and literary theory. She argues that contemporary analysis should focus on *how* some bodies and parts of bodies come to matter. As opposed to many social scientists, Butler sees constructionism not as a fixed phenomenon, but rather as "processes of reiteration," where there is not a power that acts but rather a persistent yet unstable repetition process that itself produces bodies and the categories through which bodies are seen.

For Butler, matter is "a process of materialization that stabilizes over time to produce the effect of boundary, fixity and surface" (Butler 1993: 3). In order to understand the multiple, complementary, and contradictory forces of materialization, analyses should look at how bodies register in the social world and which bodies *fail to matter*.

Sociological perspectives on the body

Critical body studies also emerges from several decades of scholarship in sociology, anthropology, philosophy, history, and other fields. Ideas and concepts developed in the second half of the 20th century laid important theoretical foundations for contemporary understandings of the body.

In the 21st century, we firmly understand the body as a socio-historical object, but this was not always the case. Sociologist Bryan Turner (1987, 1992) suggests four key developments of the 20th century that contributed to emergence of social and cultural investigations of bodies.

First, shifts in mass production and consumption from the 1920s on led to availability of cheap, durable goods (e.g., make-up, fashion) that helped to foster the body as a marker of identity. That is, the body became a way for people from various classes to use adornment to signal sexuality, gender, and class status.

Second, postmodernist theorists challenged subject/object distinctions prevalent in modernism, particularly *Cartesian dualism*, which promoted the idea of the mind as separate from the body. Postmodern theory suggests that the human body is not distinct from the self but rather deeply interrelated to identity and self-expression.

Third, feminism broadly and the women's health movement specifically were key forces that challenged predominant biomedical constructions of bodies (Ruzek 1978; Lewin and Olesen 1985). Women in the Global North rebelled against hegemonic (and damaging) medical definitions of female bodily functions, for example menstruation as dysfunction.

Fourth, French philosopher Michel Foucault's work provoked rich anthropological, sociological, and historical analysis of the social production of bodies through discipline and surveillance. Bodies increasingly became objects of state and institutional interest. Foucault's theories of social life offered a thoroughly embodied perspective on economics and politics.

For example, Foucault demonstrated that disciplinary power, employed through monitoring and surveillance, directed at bodies operates to make humans into docile and productive subjects. His work also focused beyond processes directed at individual bodies to consider biopolitics—the social practices and institutions (e.g., medicine, law, prisons, government) established to regulate a population's quality (and quantity) of life. Biopolitics and disciplinary power operate together to normalize individuals by coercing them, often by subtle mechanisms, to conform to standards and in so doing, to create self-regulating bodies and ordered nation states (see Inda 2005; Rose 2007).

Foucault's enduring contribution is evidenced by his influence on so many scholars, such as historians Thomas Laqueur (1990), discussed earlier, and Barbara Duden (1993), anthropologist Emily Martin (1989, 1994), sociologist Nikolas Rose (2006), and countless others. Foucault's work is invaluable to the field of critical body studies because his analysis demonstrates how culture constrains *and simultaneously produces* human bodies. Two critical insights thus continue to underpin critical body studies: (1) bodies experience surveillance in ordinary ways, and (2) the normalizing gaze fundamentally structures human bodily experience.

Although sociology has taken up Foucault's work, as a discipline it was somewhat late to thinking about the social significance of the body. In fact, many body scholars have suggested that the human body was largely unseen within sociology until the second half of the 20th century (Featherstone 1991; Shilling 1993; Lorber and Moore 2007). Comparing traditional to nascent industrial societies, tracing the fading import of religion, the rise of democracy, capitalism, and the emergence of the metropolis are all macro-sociological or structural questions. Given the development of the field and its early substantive foci—analyzing major historical, political, industrial, and ideological changes in Europe during the late 18th and early 19th centuries—absence of the body makes some sense in this disciplinary context.

But, as we have argued elsewhere (Casper and Moore 2009), bodies were absent from sociology not only because of disciplinary questions, but also because sociological knowledge originated through masculinist perspectives that privileged science and "reason" over lived experiences and embodiment. Historically, female bodies were framed as fleshy, leaky, and linked to the primitive and nature, whereas male bodies were associated with the mind, logic, rationality, and civility (Shildrick 1997). Sociological scholarship and practice privileged male ways of knowing and male prerogatives. Women and their bodies were understood not as central to social and political life, but rather as "other"—private rather than public, contained within the domestic sphere, and thus uninteresting sociologically.

Additionally, establishing sociology as a legitimate and distinct field of study, which was largely fueled by the efforts of Emile Durkheim and Auguste Comte, fostered definitive boundaries between sociology and the biological/psychological sciences. These historical forces effectively eclipsed any focus on individual bodies, which ironically constitute society itself. Ultimately, to position sociology as a legitimate disciplinary field, early sociologists made a strategic decision to privilege historical trends and social institutions, abstracting "the human" as a rational actor and consequently relinquishing the body and embodiment (and women) to other perspectives.

Currently, there is no unitary definition of the "sociology of the body," but sociological approaches to the body share a core set of assumptions. Sociologists reject the Cartesian mind-body dualism that detaches the self from the body, as well as theories that narrowly understand the body as merely the sum of biological functions. In other words, because we can mindfully perceive our bodies and be aware of our physical existence, some people think that the mind exists outside of the body. Instead, sociologists theorize the body as integrated with the mind and reject mind-body dualism.

Drawing from phenomenology, some sociologists focus on embodiment or the "lived body" (Williams and Bendelow 1998). This perspective seeks to highlight the material and corporeal nature of human existence by examining incarnate practices, activities, sensitivities, and habits of individual actors within their social contexts. For example, the chapters in this book enliven human anatomy by analyzing a variety of embodied experiences and representations from diverse perspectives.

Embodiment, however, remains as conceptually problematic as it is riveting. One of the most interesting aspects of this work is how awkward and difficult it is to talk (and write) about embodiment and its consequences and implications, given that little shared vocabulary exists (e.g., Martin 1989). Even when individuals share the same culture, language, social class, and ethnic background, describing the sensation of, for example, a muscle cramp and how it debilitates can be challenging.

Moving beyond a focus on embodiment, American sociologist Erving Goffman examined the body and lived experiences in terms of social structures and ideologies. Bodies are socially constrained but also something that individuals can control to varying degrees. Goffman acknowledged embodied dimensions of human agency as people attempt to manage and control their bodies. For Goffman, our ability to interact in society and to achieve desired outcomes within specific social contexts depends on the self-management of our bodies.

A key theme in Goffman's work was his treatment of bodily control and appearance as a central component in everyday encounters between people. In *The Presentation of Self in Everyday Life* (1959), he invoked a theatrical analogy to describe how agentic individuals attempt to follow cultural scripts that dictate appropriate behaviors within the presence of other people. Successful social "performances" hinge on expressive control to keep inconsistent moods and signs from disrupting the performance. In order to achieve a sense of reality or authenticity, one must master the art of "impression management," a highly nuanced technique of constant reflexive self-examination.

Goffman asserted that "the impression of reality fostered by a performance is a delicate, fragile thing that can be shattered by very minor mishaps" such as a belch, stutter, or flatulence (Goffman 1959: 56). In order to prevent embarrassment and disruption in social interactions, we must learn to manage our bodies, including their demeanor, noises, smells, waste, and facial expressions. Human agency emerges based on how we choose to act within different, often constrained circumstances. If we fail to keep our bodies in check, we risk public embarrassment and social stigmatization (Goffman 1963). Of course, some social situations allow us to relax our vigilance; for example, within families or when interacting with friends we may be able to behave in ways not acceptable in public.

French sociologist Pierre Bourdieu added another sociological dimension to studies of the body, by positing that the body is a conveyor of symbolic value that reproduces "the universe of the social structure" (Bourdieu, 1984: 193). Although his book *Distinction* did not explicitly address bodies, it took the body seriously as a repository of social values and was one of the first major sociological works to emphasize the growing importance of the body in the late 20th century.

Bourdieu places bodies within modern stratified consumer culture, arguing that the body bears the imprint of social class indicated by habitus, taste, and social location: "the body is the most indisputable materialization of class taste" in that one's hairstyle, clothing, diet, and even gait function as signs within a larger system of social positions (1984: 190). Bourdieu acknowledges that bodies are biological, yet he stresses that they are inherently unfinished, imbued with marks of social class throughout life and thus transformed.

One of Bourdieu's most significant contributions to body theory is his conception of the body as a physical form of capital. As such, the body is a resource that can be converted (by self and others) into economic, social, and especially cultural capital.

Sociologists have also theorized the body in relation to temporal shifts. For example, British sociologist Anthony Giddens conceptualized the body with respect to "late modern culture." Much of his work focused on understanding social lives and structures within the context of shifts in modernity including the late modern and postmodern periods. These shifts were characterized by a number of developments in contemporary culture including globalization, emergence of an information economy, the propensity to question scientific facts, experts, and the grand narrative (Lyotard 1984), and proliferation of imagery and simulations (Baudrillard 1994).

In *Modernity and Self-identity* (1991), Giddens argued that complex and ambiguous characteristics of late-modernity (erosion of science, changes in the family, occupational shifts, etc.) have led to increased individual reflexivity and new problems with attributing meaning to one's life. He conceptualizes the body as an "action system" within everyday life, increasingly deployed to sustain a consistent sense of self-identity (Giddens 1991: 99).

Because many contemporary individuals (especially in the Global North) can no longer rely on traditional institutional moorings (e.g., marriage, work) or roles (e.g., wife, husband, employee) to ground the self, we focus on what we know best—the body. As the self is embodied, "the reflexivity of the self extends to the body" (Giddens 1991: 77). Giddens, echoing Foucault, maintains that the contemporary reflexive body-self is continually worked on through diet, exercise regimens, implants, and so on.

Giddens's theories are useful for critical body studies in that they provide a conceptual framework for thinking about contemporary bodies and whether or not the body is experienced in significantly different ways now than in previous historical periods. Giddens raises important questions and ultimately creates a conceptual space for thinking in new ways about whether or not there is such an entity as a uniquely late-modern body.

Geopolitical bodies

The development of critical body studies requires an inherent, somewhat urgent task—seriously engaging practices of consumerism, globalization, and transnational culture. The advent of electronic media, expansion of the tourism industry, and global/transnational commerce and migration have significantly changed how we think about culture in general and thus, will have direct and indirect influence on how we think about our bodies (Appadurai 1996; Bruner 1996).

For example, research on the modern primitive body modification movement provides us with an example of how some contemporary Americans are appropriating non-Western rituals, practices, and ideologies centered on the body (Kleese 1999; Turner 1999). In addition, feminist scholars have attended to the increasing transnationalization of reproduction and reproductive bodies, including adoption (Briggs 2012), surrogacy (Twine 2011), and a host of other issues (Browner and Sargent 2011). And anthropologists have studied the pharmaceutical industry in global context, including embodied consequences for people in various parts of the world (Petryna, Lakoff and Kleinman 2006; Dumit 2012).

The contemporary moment is characterized by both transnational and neoliberal processes, which transform the experience of embodiment and recognition of particular bodies. Transnationalism focuses not merely on the "global" (as in, phenomena affecting the entire world), but rather on social, political, and economic relations and operations that transcend national boundaries. Increasingly, everything from healthcare to so-called "terror networks" to the movement of people from one country or region to another happens through nonstate formations. Bodies, especially those of the most marginal such as undocumented immigrants, are frequently caught in tensions produced by state and nonstate maneuvers.

Neoliberalism in economic terms refers to privatization, open and "free" markets, and deregulation (Harvey 2005). Used as a critical concept in body studies, neoliberalism refers to the ways in which bodies are both commodified and also reduced to their exchange value, in Marxian terms. People become a means to an end, and institutional practices and state policies under neoliberalism often emphasize the value of profit over dignity, self-determination, and social justice. Laboring bodies, reproductive bodies, and marketized bodies are key features of neoliberal landscapes.

Emergence of critical body studies

This book advances the claim that there is now a distinctive field called *critical body studies*, of which our work is a defining part and which has developed out of the perspectives discussed above. As we have shown, the body and embodiment have been part of social theory for many years now. But in many if not most of these perspectives, bodies are considered in relation to other phenomena, which are themselves framed as the central object(s) of interest. For example, in sociology bodies are dominant within the subfield of health and illness.

What is unique about critical body studies is that the body, bodies, and embodiment are centered as the *key* intellectual focus. Bodies are examined in their social, cultural, historical, economic, geographic, and political contexts; they are not ancillary to other investigations or arguments. Rather than beginning with externalities such as "society" and then moving in to look more closely at the body, critical body studies begins with bodies themselves, and asks how are they produced, distributed, marked, targeted, and otherwise "made" a key feature of social life. Feminist scholars, in particular, attend to ways the body is objectified through social relations and representations.

Critical body studies is exemplified by scholarship that examines issues of embodiment and phenomenology, "new materialism" and materiality, body projects of various kinds, performance and movement, body-technology intersections, digital bodies, body-environment connections, somatechnics, human/nonhuman relationships, and a variety of other approaches. Much of this work is about making bodies visible to social and cultural analysis while also intervening in a range of body politics, from environmental justice, to critical food studies, to reproductive justice. The work also aims to expand the kinds of bodies we see, fostering more diversity and inclusion.

Let us offer just a few examples of this new work:

In our book, *Missing Bodies: The Politics of Visibility*, we innovated a methodological approach termed the *ocular ethic* (Casper and Moore 2009), a highly useful tool for socially and culturally "dissecting" human bodies. Although there are many ways of perceiving the world, the visible is highly privileged, especially in the Global North with numerous technologies that rely on and enhance visibility. Our ocular ethic is an empirical strategy designed to reveal hidden or ineffable aspects of human life. Dimensions of the ocular ethic include *focusing, magnification*, and *visualizing*.

By focusing, we mean the processes of drawing our attention and analytic gaze to often-marginalized bodies, individuals, and groups in social life. For example, in *Missing Bodies* we analyzed "African AIDS" and the erasure of sick African bodies in the service of U.S. national security concerns. The act of visualizing these bodies, of critically focusing on the actual embodied experience of this trend, is, we argue, an intellectual and ethical responsibility.

The ocular ethic also uses techniques of magnification, including ethnography and cultural analysis, to reveal, resituate, and recuperate bodies in order to examine the broader contexts that shape and constrain them. Ocular lenses are devices that can magnify the image formed by the "objective" lens, and this then can focus our vision on a particular aspect of images in a photograph or tableau.

Just as the magnification of a microscope can be increased up to a limit of resolution by changing the objective lens, we practice an ocular ethic of magnifying bodies and body parts that are hidden or concealed, sometimes deliberately so. In *Missing Bodies*, we explored the consequences for marginalized people of Hurricane Katrina, as the bodies of the aged, disabled, and people of color were relegated to zones of neglect and even violence. Here, magnifying those bodies tells an important story about race, class, ability, and "natural" disasters.

The interdisciplinary area of *somatechnics* begins with the provocative suggestion that bodies are always already technological; that is, corporeality is understood to be historically "bound up with" technologies, techniques, and technics. Institutionalized through the journal *Somatechnics*, a series of international conferences, and the transnational, interdisciplinary Somatechnics Research Network, this area of study is committed to offering "ethico-political consideration of a wide range of practices and techniques."

Another area of study within critical body studies is *new materialism*, exemplified by the work of scholars such as Karen Barad (2007), Mel Chen (2012), and Jane Bennett (2010). New materialism reflects a serious engagement with the relationship of matter to social and cultural interpretations. Matter is something distinct from our thoughts and sacred meanings; it is commonly defined as something that occupies space and has mass. Theorists in this area explore the ontology, or essence of being, of matter as deeply consequential for how we come to know our world and constitute it. And yet, several scholars invested in the study of materialism and materiality have raised questions about just how "new" this new materialism is.

In sum, critical body studies works to center the body, bodies, and embodiment, marking these as fundamental to the production of knowledge and to social life. Social and cultural analysis begins with the body—including its "biology" and "sociality"—and works outward, to amplify the social operations that give it shape and meaning.

Several historical and contemporary theoretical strands and perspectives overlap and intersect to construct this approach to bodies. As critical body studies continues to develop, the field will undoubtedly morph in new directions, with tentacles reaching into as-yet undeveloped areas of scholarship.

Related interdisciplinary fields

It is important to note that critical body studies is an interdisciplinary endeavor. Although we are grounded in both sociology and women's studies, our work—like the work of many critical scholars of the body—intersects with other fields, such as animal studies and reproductive studies.

Here, we briefly discuss some of the major fields that have begun to incorporate and engage with critical body studies. We do so in order to help readers situate the material presented in this book and also to organize and interpret their own engagements with the field.

Affect studies

Sometimes called affect theory, affect studies attends to sensation, or rather the embodied, expressed experiences of feeling and potentiality. Scholar Russ Leo (2011) defines affect as "indices of experience, attempting to describe the sense of the world in detail, with sharper attention to the rich lived realities of seemingly sterile concepts." Not quite feeling, not quite emotion, and not quite expression, affect is a difficult concept to pin down. In *The Affect Theory Reader*, Gregory J. Seigworth and Melissa Gregg (2010) note that: "affect arises in the midst of *in-between-ness*: in the capacities to act and be acted upon."

Relevant to critical body studies, affect is said to be located "in those resonances that circulate about, between, and sometimes stick to bodies and worlds" (Seigworth and Gregg 2010: 1). It is, then, something ineffable but remarkably present in the spaces, movements, and encounters between and among bodies and in bodily experiences. Words such as network, shimmer, contagion, mimesis, virtual, desire, pleasure, and movement are staples of affect studies, which tends to characterize the body as "biomediated" (Clough 2010). Affect theory can be firmly located as an intellectual project of the 21st century, although its roots extend from post-Cartesian philosopher Spinoza.

Ethnic studies

Broadly, ethnic studies looks at race and ethnicity as they are socially constructed and relied on as the basis for social organization, including formations and practices of inequality. Ethnic studies emerged as a response to the eurocentrism of mainstream disciplines such as sociology and anthropology, and also grew out of civil rights activism of minority groups in the Global North. Ethnic studies scholars frequently analyze race, gender, class, and sexuality alongside ethnicity, or what feminist legal scholar Kimberlé W. Crenshaw (1991) productively called "intersectionality."

Comprehensive ethnic studies departments often encompass multiple perspectives and groups, while specific programs have also emerged. In the United States, these include African American and Africana studies, Latino/a and Chicano/a studies, Native American studies, and Asian American studies. In Canada, the ethnic landscape looks quite different, with an academic focus on indigenous communities as well as, for example, historic tension between English and French communities in Québec. In Australia, ethnic studies includes a focus on Aboriginal people and Torres Strait Islanders, critiques of whiteness, as well as historical examination of white settlers/invaders. Scandinavian countries, such as Norway and Sweden, are increasingly focused on ethnic studies as their demographics change with new immigrants, particularly from Africa.

Wherever the field is found, ethnic studies is a deeply political project, dedicated to understanding and eradicating white supremacy through critical pedagogy, legal and policy analysis, and relevant social scientific research. Ethnic studies typically calls for an intersectional approach, offering a critical perspective that explores the interrelationships of multiple systems of oppression.

Ethnicity often serves as a cultural guidepost for bodily interventions, such as tattooing among Maori in New Zealand. Yet ethnicity may also be the basis for "ethnic cleansing," genocidal practices wrought on bodies through mutilation and death.

Gender and sexuality studies

Gender and sexuality studies, also sometimes called gender and women's studies or feminist studies, is the interdisciplinary study of sex, sexuality, and gender, particularly how and with what consequences people are categorized and stratified into "female" and "male." The field is committed to investigating how the dual meanings of "sex"—as a classification system that enables gender norms and as a designation of intimate desire and action—reverberate to put gender and sexuality in very close proximity to each other.

The field examines, among other topics, how "biological" characteristics are used as a basis for social organization, and it attends to inequalities between men and women. Included in gender studies are investigations of gender representation, gendered performances, the social construction of gender and sex, sexuality and its expressions, gender in social institutions, politics and policy related to gender, transnational feminisms, LGBTQI identities, and various social processes in their gendered dynamics such as socialization, reproduction, education, migration, and employment.

Many (but not all) scholars of gender and sexuality studies are feminist in their political and intellectual orientation, and many (but not all) work against racist and imperialist structures and practices. The field has been greatly influenced by the growth of queer theory since the 1980s (Jagose 1996). The body is relevant in gender and sexuality studies particularly as it is a target of state intervention and subject to various kinds of violence.

Transgender studies

Transgender studies is a deeply interdisciplinary field of research, theory and pedagogy, influenced by gender and sexuality studies, LGBT activism, and queer theory. By the 1990s, a burgeoning body of trans-historical and cross-cultural literary, documentary, performance, political, and anthropological work had developed into the new field of transgender studies (see for example, Currah, Juang and Minter 2006; Stryker and Whittle 2006; Stryker and Aizura 2013). This new field linked insights and analyses drawn from the experience or study of transgender phenomena with the central disciplinary concerns of contemporary humanities and social science research.

The meaning of "transgender" itself has shifted significantly since the word first began appearing in cross-dresser community publications in the late 1960s. Political theorist Paisley Currah suggests that, since the early 1990s, the term "transgender" has become most commonly used to describe people, in the U.S. and elsewhere, whose gender identity or gender expression does not conform to social expectations for their birth sex (Currah 2006).

Historian Susan Stryker (2008:1) defines transgendered people as those:

> . . . who move away from the gender they were assigned at birth, people who cross over (trans-) the boundaries constructed by their culture to define and contain that gender. Some people move away from their birth-assigned gender because they feel strongly that they properly belong to another gender in which it would be better for them to live; others want to strike out toward some new location, some space not yet clearly defined or concretely occupied; still others simply feel the need to get away from the conventional expectations bound up with gender that were initially put upon them.

Disability studies

The field of disability studies emerged, in part, out of the disability rights movement, as activists and advocates struggled to make societies more accessible to people with disabilities while simultaneously locating disability not in individual bodies but in the environments around them. Intellectually, disability studies has worked to challenge meanings and definitions of "normal" and "able" with respect to the body.

Some work has situated people with disabilities as a minority group, subject to discriminatory practices much as any other minority group and in need of similar legal remedies. Other work has challenged the medical model of disability with its core notion of pathological or impaired bodies. And still other scholarship has advanced the idea that what we know as "disability" has been constructed through discourses and practices such as literature, visual representations, law and policy, interpersonal relationships, technologies, education, and politics.

The role of the body within disability studies is complicated. While some disability studies scholarship has moved away from a focus on embodiment in order to center the individual as a subject of rights, other scholarship focuses more concretely on embodiment and materiality as a way to make sense of people's lived experiences, including pain.

Critical health studies

Critical studies of health are not only interdisciplinary, they are also diverse ranging from medical sociology to medical anthropology to community and urban studies. Each has its own set of theoretical perspectives, but what they all share is a commitment to locating health and illness in social contexts. They also critique the biomedical model, which they perceive as reductionist.

Medical sociology—one of our fields of expertise—is diverse, with a branch of study focused on political economy of health that rarely takes the body into consideration. Sociology of health and illness, attentive to lived experiences of pain, illness, and suffering, tends more toward integrating the body into scholarly analysis. Often in this work, embodiment and materiality (e.g., the physical dimensions of illness), are conceptualized as the terrain on which illness unfolds, or as a context for suffering.

More topically, from the 1980s onward, research on HIV/AIDS increasingly assumed a critical focus. For example, sociologist Cindy Patton (1990) forged a new way of looking at disease that was both very interdisciplinary and politically engaged. Significant work on HIV/AIDS from scholars in sociology, anthropology, literary studies, cultural studies, and other fields followed, marking the emergence of HIV/AIDS as not only a signal moment in global health, but also as an impetus to intellectual evolution.

Aging studies

Everybody ages, from birth through death; this is a basic fact of human life. The study of aging is most commonly known as gerontology, but scholars have begun to use the more inclusive and interdisciplinary term "aging studies."

Researchers in this field investigate various aspects of aging, including social, psychological, and physiological dimensions. The aging body is a central object in this field, as scholars from a variety of backgrounds study such topics as how bodies become frail, brain function in the very old, the beneficial psychological effects of movement, loneliness among elderly people who have lost spouses and friends, and a range of other issues.

Only recently has scholarship in gerontology begun to reflect insights from critical body studies. For example, borrowing from disability studies, gerontologists have devoted more attention to social context than to individual characteristics such as frailty and psychological distress, focusing instead on how environments can profoundly shape the embodied experiences of aging.

Sport studies

Sports—both local and global, both amateur and professional—are popular worldwide. In many societies, they are big business, too, ranging from professional football, basketball, and baseball in the United States, to soccer throughout much of the world, to cricket in England and its former colonies. Sport studies is an interdisciplinary field drawing from sociology, anthropology, history, kinesiology, and cultural studies that takes "sport" as an object of analysis.

The body is central in sport studies, as scholars investigate the gender, racial, class, and performance dynamics of sporting bodies and practices. For example, an analysis of South African track star and murder suspect Oscar Pistorius might focus on disability and masculinity as they are enacted by his prosthesis. A cultural account of tennis stars Venus and Serena Williams might start with their highly visible role as African American women athletes in a sport long dominated by whites. Theresa Runstedtler's 2012 book, *Jack Johnson, Rebel Sojourner*, examines the African American boxer's life in relation to the history of U.S. race relations.

Thus, in sport studies the sporting body and its meanings is as significant as the sporting activity.

Security studies

Influenced by Foucault, post-9/11 "security studies" is an emergent interdisciplinary field that examines the intensification of social control through the use of techniques of visibility and internalization of control directed at bodies. State-sanctioned institutions refine techniques to see and observe the movements and behaviors of bodies through biometrics and biotelemetrics. An excellent example of this is the apparatus of airport security.

Surveillance cameras exercise extraordinary power by monitoring potential criminal activity, maintaining security, and controlling anything deemed to be "deviant" according to the ideology of those in power. Although contemporary security and surveillance activities may be experienced as unprecedented, the state has a long history of making bodies legible, both individually and in the aggregate.

Social scientists Jane Caplan and John Torpey (2001: 8) track "the 19th-century development of documentary practices through which every citizen, not just the delinquent or deviant, was to be made visible to the state: not by physical marks on the body, but by the indirect means of registrations, passes, censuses, and the like."

Although security studies is not always in full engagement with critical body studies, it is abundantly clear that the field of security and surveillance studies would not, in fact, exist without a focus on the body. From airport security to cameras in urban areas, from parole bracelets to incarceration and border patrol, the body is relentlessly the target and focus of security measures.

Although we do not have space to focus on them here, other fields, such as food studies, migration studies, and fat studies, also increasingly intersect with critical body studies. We touch on these in some of the book's chapters.

Core concepts in critical body studies

The success of critical bodies studies as an interdisciplinary specialization and field has meant there is a growing language, including terminologies and concepts, used to communicate significant ideas and theories. We provide working definitions of these concepts in the book, which will reappear throughout individual chapters.

Affect: emotional or cognitive impact, the experience of emotion or feeling. Affect can also indicate the pre-conscious level of sensation that is often nonverbal and sometimes subconscious. Affect is important because there are somatic presentations of affective changes — such as hair raising on the back of the neck, sweating, and increased salivary production.

Biopolitics: a term most significantly developed by Michel Foucault, which describes the ways the state or governing powers become increasingly concerned with the population's health and bodily wellness. Biopolitical processes are signified by a growth in the state's surveillance of health aggregates — birth rates, death rates, morbidity statistics, and immunization statistics.

Body projects: a term innovated by historian Joan Jacobs Brumberg (1998) to explain the ways we are encouraged or coerced into seeing our body and its parts as infinitely perfectible. These body projects drive our consumptive practices toward the purchase of pharmaceuticals, cosmetics, body treatments, surgeries, health regimens, lifestyles, and ideologies.

Corporeality: having materiality, or a carnal, fleshy, physical form. Corporeality is the opposite of spiritual or metaphysical; the somatic, material, and physical are tantamount. In body studies, corporeal is often used interchangeably with "material" and both are understood as the opposite of theoretical or abstract.

Cultural capital: a term emerging from Pierre Bourdieu's scholarship. The transmission of norms, values, attitudes, and knowledge through social practices that increase the possibility of success from one generation to the next. These are noneconomic goods that include verbal skills or aesthetic preferences that one portrays as membership in a certain strata or class. For example, knowing when and how to use certain utensils for eating is a form of cultural capital that can either impress or offend one's hosts.

Disciplinary power: a term originating from Michel Foucault's examination of bodily rituals used for monitoring, training, and socializing of bodies for specific tasks and roles. Disciplinary power regulates the behavior of individuals as members of the social body through social institutions including the medical industrial complex, educational structures and discourses and the media. Disciplinary power is executed through the repetitive drilling and surveillance of bodies to create docile and predictable citizens.

Embodiment: the quality of having a body and perceiving and being in the world through the body. Being embodied is often taken for granted, until the body fails or is somehow disrupted. Sociologists examine these very taken-for-granted embodiments through methods such as ethnomethodology.

Habitus: a concept originally outlined by Marcel Mauss and then expanded by Pierre Bourdieu. The term is used to describe the unconscious patterns of doing, thinking, speaking, and perceiving which an individual embodies and displays in the social world. One's gait can be considered part of the habitus. Habitus can also transmit a certain social status.

Impression management: Erving Goffman coined this term to describe internal practices of managing external expectations. In other words, how one comports the body, displays emotion, and reacts to social situations is invariably done in light of perceived social reception.

Inscription: literally or symbolically engraving or writing into the body. Bodies are often inscribed with culturally mandated or personally chosen adornments or markings. Inscription also indicates that cultural norms or taboos are symbolically written into bodies as a means of socializing people.

Laboring bodies: bodies that work or perform, typically in an economic setting; laboring bodies both produce capital and are subject to capital as workplace innovations target body function for the 21st century; laboring bodies are antithetical to bodies at rest or leisure.

Materiality: consisting of matter, tangible. Bodies are material and we use materiality to note that there are "flesh and bone" potentials and limitations of bodies. Bodies are also representational and notional, as well as ideological and discursive. In other words, there are actual physical, corporeal bodies as well as ideas about what bodies are or should be. We use materiality to designate, in pragmatic terms, the body's organic matter.

Morphology: the study of the shape and form of the body. This term indicates that the body is moldable and flexible, and as a material substance and symbolic projection can contort and twist itself into different shapes. The morphology of the body is important, as there is a limit to the shapes into which we can contort ourselves, and yet we are constantly, as a species, working to test the limits of our corporeality through transformation.

Neoliberalism: a shift in social relations that is characterized by governmental and economic institutions collaborating to create opportunities that favor growth in the service of large global and transnational enterprises. This shift has obvious and tacit effects on economic and social relations. As social theorist David Harvey (2007) has suggested, neoliberalism produces the doctrine of market exchange and profit as an ethic guiding human action.

Normalization: characterized by the process of conforming to prevailing social expectations of the ruling class; when individuals work to come in line with social norms through adherence to socially expected practices. There are vast social pressures on the body to comply with normalized practices with respect to notions of health, practices of sex and signs of sexuality, gender displays, body size, and ability.

Somatic: of the body. That is, a characteristic of the physical body that often concerns the survival or comfort needs of the body. Noting that we are working at the level of the somatic means that we are talking about materiality, corporeality, physicality, and morphology.

Stigma: refers to the pejorative labeling of a characteristic or practice of another individual. Erving Goffman used the term to explain the distortion and discrediting of an individual in a social group based on the group's designation of the individual as unworthy or flawed. People with AIDS whose bodies are literally physically marked with bruises as well as symbolically marked as "diseased" are stigmatized. Stigma often includes the exclusion of individuals from everyday practices of social life.

An owner's manual

This book is written for advanced undergraduate students in the social sciences, humanities, and allied health fields—and, of course, their teachers. We have included several images alongside the text to illustrate the myriad ways the body is figured and represented in contemporary cultures.

As scholars, our own research has been primarily conducted within and on North America, and most often takes place in the United States. Given our geographic and structural location, in the absence of firm evidence we are reluctant in this book to speak for or represent people and issues from other countries. Monica has worked on health issues in Canada, at the U.S. Mexico border, and in Puerto Rico, New Zealand, and the South Pacific, although we no doubt have only partial comprehension of local and regional experiences outside the U.S. *The Body* is thus somewhat more "U.S.-centric" than would be ideal. Whenever possible and relevant, we rely on transnational scholarship throughout this text so as to augment the U.S. data and case studies.

Further, while existing labels are problematic when referring to stratified global differences, we use the terms *Global North* and *Global South* to indicate socio-economic and political divisions and hierarchies. The Global North typically includes North America, Europe, and "more developed" parts of Asia, whereas the Global South refers to Africa, South and Central America, and "less developed" parts of Asia. Taking our lead from DAWN (Development Alternatives with Women for a New Era), the term "south" is a political and ideological location-position that describes poor and marginalized people's experiences. Thus, we prefer this terminology to the language of "development," which carries with it meanings of value, attainment, and hierarchy.

The following chapters are grouped within three sections: Building Blocks, Systems and Their Representations, and What Others See. The chapters themselves follow a similar structural organization with component parts. These are a brief introduction to the subject including a biological and sociological overview, key concepts and terminology, two "A closer look" substantive explorations (or case studies) for every chapter, and five discussion questions. "A closer look" is meant to provide readers with more in-depth and provocative examples of social and cultural interpretations of bodies, body parts, and body systems. These sections are shaded throughout to distinguish them visually in the chapter.

We have written and organized *The Body* so that it flows relatively seamlessly from beginning to end, but also so that it is not necessary to read the chapters in order. Teachers and students can pick and choose, and can also read the chapters along with other materials that focus on each topic.

Learning objectives of the book

- Students will gain familiarity with key concepts in critical body studies, including affect, biopolitics, body projects, corporeality, cultural capital, disciplinary power, embodiment, habitus, impression management, inscription, laboring bodies, materiality, morphology, neoliberalism, normalization, somatic, and stigma.
- Students will be able to compare and contrast important terms, and to be able to apply them to substantive issues and case studies.
- Students will develop the ability to integrate sociological perspectives of the body with sociological theories of inequality based on race, class, gender, sexuality, ability, age, health, geography, and citizenship status.
- Students will be able to analyze and interpret a range of human bodily experiences, biomedical and other interventions, and embodied practices.
- Students will be able to identify key ideas that differentiate social and cultural perspectives on the body from standard biological and biomedical framings.
- Students will learn to recognize and be able to identify key theorists and social and cultural researchers of the body.
- Students will learn to think critically and analytically, including the ability to use core concepts from critical body studies in their written work.
- Students will develop the ability to compare and contrast bodily representations and practices from different parts of the world.

PART I

Building Blocks

chapter 2

Genes

Introduction

This chapter explores genes, considered the basic "building blocks" of life. We begin with a brief look at new research in human genetic engineering, including manipulation of the human genome that pushes at ethical boundaries of acceptable interventions. We turn next to physiological and cultural understandings of genes. We discuss genetic determinism as well as its counter, social constructionism, which has been used to explain the rise of genetic explanations.

Addressing the history of eugenics, the chapter also examines how race, criminality, and sexual orientation have been constructed through a genetic lens. The rise of genetic explanations is deeply connected to meanings of heredity, and these issues are explored through an examination of decisions about reproduction. The chapter concludes with a discussion of the ways in which biological metaphors, in this case regarding genetics, can come to overshadow other explanations and meanings of human life.

Brave new worlds

Genetic engineering refers to the manipulation of the genetic material of any organism. Historically, such tinkering has been routine in laboratories and classrooms, and many people engage in these practices. But what would it mean if scientists could alter *human* DNA in ways that permanently affect future generations? Although *human genetic engineering* has long been "off limits," according to common ethical understandings, in October 2012 geneticists in Oregon replaced so-called "defective" material in human eggs with "healthy" DNA provided by donors.

The aim of the experiment—to prevent genetic disease in future generations—is certainly noble. But the research opens up a Pandora's box of ethical dilemmas, including mistakes in genetic manipulation that might be passed to offspring, the specter of engineering "superior" human beings, and the challenges to human identity that might be posed by having more than two genetic parents.

Responses to the Oregon research—and to most human genetic engineering practices—invariably invoke Aldous Huxley's 1932 novel, *Brave New World*. The novel foreshadowed many 20th and 21st century technological "advances" in human reproductive and genetic sciences. Required reading in the Western literary canon, the book echoes (and likely fuels) collective fears of the loss of human values to ever-expanding consumerism as humans quest for "perfect" offspring. In the span of half a century, we have witnessed *in vitro* fertilization, surrogacy, sperm donation, egg freezing, genetic diagnosis, prenatal diagnosis and treatment, the Human Genome Project, and human genetic engineering.

A common theme in ethical commentary is that technical "advances" move far more quickly than society can comprehend and adapt to them. The "brave new worlds" created through manipulation of the human gene pool are scary and unfamiliar, evoking both anxiety and excitement. What's more, these new worlds have the potential to be less liberatory and more likely to reinforce already existing social inequalities (Bliss 2012).

Sociologist Troy Duster's (1990) pioneering study in *Backdoor to Eugenics* was among the first to analyze social consequences of genetic research and helped to spur development of the field of *genethics*. Duster showed how contemporary genetics practices reaffirm racial inequalities, but because this happens through the so-called "objective" language and practice of science ("the backdoor"), the eugenic politics shaping the work are occluded.

Our genes, ourselves

"No blood test. Just a cheek swab," advertises DNA Spectrum, one of a handful of private genomic scanning services available for testing one's DNA profile including genetic risk for 47 diseases (http://www.dnaspectrum.com/). Individuals are able, for between US$100–$500, to purchase information about their particular genetic make-up. Direct-to-consumer tests claim to give percentages of likelihood for conditions such as Alzheimer's, baldness, and rheumatoid arthritis.

We can test ourselves, our partners, our children, our embryos, our gametes, and even our pets and livestock. Armed with this new genetic information, the story goes, we can make more informed decisions about how to navigate (and possibly treat) the dangers lurking inside our bodies.

But what is DNA? And how does it connect to genes?

DNA, or deoxyribonucleic acid, is material in living cells that provides the template for its own duplication. Originally defined as the complete directions, blueprints, or codes for particular proteins, a gene is a segment of DNA that carries a specific set of instructions to perform (or regulate) a particular task in relation to cell assembly or functioning. In human populations, genes are transmitted across generations through sexual reproduction; they contain hereditary information.

Over the past few decades, the use of genetic codes in how we think about the world has dramatically expanded. In fact, it is difficult to disentangle the scientific description of what a gene is from the ways in which genes have entered our collective knowledge as the key to. . .everything. Indeed, as sociologist Dorothy Nelkin and historian Susan Lindee (1995) persuasively argued, the gene is the preeminent icon being deployed in everyday life to explain everything from homosexuality to alcoholism, obesity to IQ, voting behaviors to marital choices.

Genetic explanations for phenotypic attributes such as eye and hair color evolved from Mendelian genetics of the 19th century. Gregor Mendel's experiments on plant hybridization led to theories of inheritance, and Mendel is widely considered to be the founder of genetics as a field. The idea that features such as eye color, hair color, and height are genetic is largely uncontested and, even more, popularized.

But genetic explanations for other types of human expressions, such as IQ and behavior (Jasny, Kellner, and Pennisi 2008), are much more controversial.

Theorizing nature and nurture

Contention about underlying causes of human behavior and social action is often understood in terms of the nature/nurture debate. That is, are we "programmed" to behave in certain ways, or do we learn how to behave as members of a social community? Within the social sciences and cultural studies, there are ongoing, significant conversations and disagreements about paradigms that purport to explain human actions, identities, and behaviors.

Possibly the most common paradigm is *biological essentialism*, which claims that many (even most) human behaviors and attributes, such as nurturance, aggression, sexuality, and altruism, are biologically determined. That is, we are "hard-wired" for everything from eye color to criminality. An example of this paradigm is when mothers are said naturally to care for their children, with "maternal instinct" stemming from our inherent programming as a species. This claim implies that humans are not social beings who learn through socialization, but rather that we perform tasks and create relationships because of biological drives or determinisms.

During the 1970s, *sociobiology*, or the study of the biological basis for human behavior, was quite popular. Sociobiology, popularized by Edward O. Wilson (1975), explains there are genetic imperatives that have been inherited through generations, and behavior is thus an effort to promote one's genes. Evolutionary biologist Richard Dawkins (1976) goes further: "We are survival machines – robot vehicles blindly programmed to preserve the selfish molecules known as genes. This is a truth which still fills me with astonishment."

According to these theories, genes are destiny and social life is incidental to evolutionary biology. Yet within a social constructionist perspective, the categories of what is "crime," "nurturance," "sexual orientation," and so on are historically created and reproduced in specific contexts. Bodies are slotted into categories that are already established and that help to structure scientific knowledge about bodies, as discussed next.

Genetic determinism, akin to biological essentialism, suggests that genes have ultimate causal potency and thus trump every other explanation for human differences in social success, disease distribution, gender, race, sexual orientation, and so on. While research has shown that some diseases are genetically linked, the paradigm has shifted from identification of specific genetic mutations and conditions to *tendencies* toward illnesses and behaviors.

This shift has consequences for how we approach genetic "abnormality" —we have moved from targeted treatment of specific conditions to earlier diagnostic and preventive interventions, some of them applied prenatally to embryos in the womb (Pray 2008). With this shift have come new categories of risk, for example women at genetic risk for breast cancer or Maori at greater risk for violence (Hook 2009).

In the case of the Maori having a "warrior gene," several geneticists and social scientists have forcefully argued that the original researchers overreached their data and fueled reductionist arguments about violence and race in ways that did not consider socio-economic as well as racial stereotypes in explanatory modeling. The discourse of genetic risk is therefore rife with preexisting socio-cultural prejudices that must be examined.

Although there are strict adherents to the doctrine of genetic causality, new research considers biosocial correlates of disease. Environmental triggers, such as exposure to toxins or psychosocial stress, interact with genetic mutations and variations to lead to certain types of cancer, heart disease, or chronic conditions. This interaction of genetic factors with social conditions, or gene-environment interaction, creates the conditions for individuals' health and wellness.

Contrary to views that express biological determinism is *social constructionism*, a theoretical paradigm that suggests much of what humans do, we do because we have learned to do so from other humans beings. This perspective relies on historical, sociological, and anthropological meanings and categories to explain social life. Within this framework, understanding, for example, the division of labor, stratified organization of cultural groups, inequality, and use of language are paramount to understanding the systems and meanings that organize human life.

While social constructionists accept there are material realities in the universe and that bodies and nature may be obdurate, they argue it is impossible to perceive and communicate about these materialities without frames of reference or prefabricated interpretations and behaviors we have been taught (Mead 1934; Berger and Luckmann 1967).

Sociologists are typically very quick to establish the constructed features of social life, in part because biological essentialism's explanation of why humans do what they do has so predominated. There is a sense that, in order to counter the prevailing wisdom about how humans act, as natural or genetic, sociologists must shy away from engaging deeply with genetic explanations—and instead often criticize this model (Udry 1995).

For example, within social constructionism, a mother is understood to nurture her children because of deeply regulated norms and rewards for strict adherence to gendered maternal behaviors of previous generations, and not because of the genetic code, hormonal drives, or physiological imperatives (1994). Social understandings of motherhood are, within this perspective, better able to account for cross-cultural and other differences in how women mother.

Yet mothering, like so many behaviors, is more complex than the dichotomy of "biological essentialism" and "social constructionism" allows. Our longstanding theories and methods, particularly those reliant on dichotomies and binaries of *either/or*, may not be adequate to the task of revealing connections between "natural" tendencies and the things we learn through "nurture."

Recent scholarship, such as feminist accounts of primatology (Haraway 1989), the new materialism (Coole and Frost 2010), and critical body studies, offers a hybrid understanding of the ways that human behavior—including embodied practices—may be shaped by intersections of the so-called "biological" and the "social."

A CLOSER LOOK

Genetic discrimination

Treating an individual differently because of their perceived genetic array (including mutations) is known as *genetic discrimination*. This attitude might seem farfetched, especially because we do not all know our genetic make-up. The 1997 film *Gattaca* is still an impressive, memorable account of a world in which genetic discrimination determines the decisions and opportunities of a person's life.

There is some reason to be suspicious of how genetic information might be made available to employers or health care providers and insurers. In the United States, preexisting conditions used to deny insurance coverage for individuals take on a whole new meaning if genetic information is available. In the UK, the Human Genetics Commission has created a Genetic Discrimination Monitoring Group to track and assess the use of genetic information in employment and insurance.

Prejudicial feelings about individuals perceived to be at risk for illnesses based on heredity create a new status by which to judge others—the genetic status—and establish a new "normal" gene against which to compare people. The 2003 International Declaration on Human Genetic Data of UNESCO, the most important international instrument on genetic privacy, outlines guidelines for consent, collection, processing, use, and storage of genetic data (Motoc 2008).

Eugenics, or social engineering

Eugenics is the practice of "improving" a population or species by selective breeding or genetic engineering. Any discussion of human genes requires acknowledgement of historical and contemporary attempts to control propagation of the species, including the use of ideas from the natural sciences to guide policy decisions. The term "eugenics" comes from Francis Galton (1822–1911), a cousin of Charles Darwin who also first used the term "nature versus nurture" (Moore 2003).

Since Galton's time, eugenics has had many expressions in many different social contexts. But in general, through eugenic practices, people in societies with more power have attempted to breed out "bad" traits and breed in "good" ones through genocide, forced reproduction, sterilization, and other violence against those less powerful (Stern 2005). Often, these social engineering efforts are pursued in tandem with scientists and clinicians.

Eugenics can be viewed along a continuum of practices. Scholars speak of *positive eugenics*, which refers to encouraging reproduction among genetically desirable people. *Negative eugenics* refers to removal of genetically undesirable people from the population, through extermination and genocide, sterilization, and other technologies.

Prenatal care, preconception care, contraception, abortion—are all, in part, about controlling the ways in which humans reproduce while also controlling *who* reproduces. For example, who has access to contraception? Which nations promote more births (pronatalism) and which promote fewer births (antinatalism)?

China, with its one-child policy, is an excellent example of an antinatal state (Greenhalgh 2008). As sociologist Gail Kligman (1998) illustrated, Ceausescu's Romania, with its repressive anti-abortion practices designed to encourage reproduction, reflected a pronatal position.

In both cases, it is pregnant women's bodies on which these policies are enacted.

FIGURE 2.1 American Eugenics Society
Photo courtesy of the American Philosophical Society—"Eugenic and Health Exhibit, Fitter Families exhibit and examination building, Kansas State Free Fair, 1920"

The concept of *Social Darwinism*, intimately linked to eugenics, emerged in the early 1900s and applied notions of species survival and breeding to encourage hierarchical classifications of groups and individuals into categories of "fit" and "unfit." Social Darwinism as an ideology has been used to justify imperialist, nationalist, and racist policies (Claeys 2000).

Germany's Third Reich notoriously schemed to create a master race through reproductive and extermination policies. The Nazis instituted several programs to eliminate "undesirable" people (most notably, but not only, Jews), in a concerted effort to create Aryan white supremacy (Proctor 1990). Here, eugenics took expression in the Holocaust, coupling scientific practices and medicine with state-sponsored violence.

Genetics, race, and difference

Contemporary interest in human genetics led to large-scale scientific research projects, many of which have posed numerous ethical and legal concerns (Kevles and Hood 1993). The Human Genome Project, or HGP, aimed to identify and make accessible for research studies the 20,000–25,000 genes in human DNA.

In 2003, the 13-year collaboration between private and public organizations in the United States, the United Kingdom, Brazil, Japan, France, Germany, China, Australia, and others was completed. The full sequence is available at http://www.ncbi.nlm.nih.gov/genbank/. The HGP has been billed as an opportunity to discover, research, and treat biological and increasingly psychological and social disorders. It has also provoked ethical debate about the aims of research, the long reach of eugenics, and corporatization of science.

Importantly, the HGP has proven how incredibly *similar* the human species is: "Genomic researchers have found that any two human individuals are approximately 99.9% the same genetically, and it is hypothesized that the most important genetic material for human functioning is encompassed in that shared set. The 0.1% difference, although comparatively small, represents about three million differences between individuals' DNA. . .No sharp genetic boundaries can be drawn between human population groups" (Bonham, Warshauer-Baker, and Collins 2005: 11).

And yet, despite this finding from the HGP, human beings continue to be represented as profoundly different, and to organize society (including genetic testing, treatment, and behavioral studies) along these differences, which are perceived as "racial," "ethnic," and/or "cultural" (Hammonds and Herzig 2009; Roberts 2012).

Race as a category is a social creation that changes across time and space, and it means different things in different parts of the world. It is used in *social classification* systems to organize individuals into groups based on appearance, customs, behaviors, and geographical ancestry. These groups are often stratified, with hierarchies reflecting the interests of the dominant group. As anthropologist Clarence Gravlee (2009: 54) argues, "systemic racism becomes embodied in the biology of racialized groups and individuals, and embodied inequalities reinforce a racialized understanding of human biology."

Sociologist Ann Morning (2011) studied how Americans think about race. She interviewed scholars and college students and examined high-school scientific textbooks, and she also looked at how companies use genetic information in ways that perpetuate biological essentialism. For example, some companies market drugs on the basis of race. She found that understandings of race are deeply tied to genetic meanings, even while scholars may disagree about how "social" or "natural" race is.

Similarly, sociologist Johnny E. Williams (2011) examined the rise of genetic racial differences, seeking to "uncover the ways in which human subjectivity (i.e., racialized social contexts, ideas, vocabularies, and distinctions) imposed on genome variation reality become objective fact" (554). His project unveils how scientists have shifted their analysis from the genetics of difference to the genetics of *racial* difference. Through a careful analysis of genetic research that found *no* group racial differences Williams shows how, despite lack of evidence, media reports suggest racial differences are genetic rather than cultural or social. He argues, in fact, that racial ideology is encoded in genome variation research; that is, racial categories shape the research itself, including the kinds of questions asked and about whom.

Michael Omi and Howard Winant (1986) innovated the term *racialization* to describe the application of ideas and meanings of race, especially racial difference, to social practices. They suggest that application of racial categories is not static, but rather that the making of race is an ongoing process. Not only are genes themselves racialized—that is, understood through a lens of racial difference and inequality—but genetic meanings are enrolled in other racialization projects, and increasingly are used to create and sustain them.

For example, racialized genetic explanations have become important in health care, especially regarding targeted interventions such as BiDil, which is used to treat heart failure among African Americans. As anthropologist Jonathan Inda (2011) shows, racialized interventions may foster discrimination by suggesting a genetic basis for health. But they may also foster notions of *biological citizenship*, or a sense of community based on shared biological traits or experiences, that can lead to health activism and new social formations.

Above all, racialization is a social process that affects other meanings and practices.

"Criminality"

One of the most fraught areas of genetic research has to do with criminal behavior, or an innate perceived "criminality" among some people that ostensibly leads them to offend. Some research has looked at whether genes influence an individual's propensity to crime and, if so, how.

One study by a criminologist in Texas that used data from the National Longitudinal Study of Adolescent Health found that genes could influence a person's pathway toward a life of crime. The author noted that "there is no gene for criminal behavior," but also stated "there are likely to be hundreds, if not thousands, of genes that will incrementally increase your likelihood of being involved in a crime." As with health conditions and susceptibility, here too we are in the murky realm of *genetic predisposition*. Genetic predispositions tend to be understood as associations rather than direct causal links between genotype and phenotype, creating interpretive space for linking genes to behavior.

Unfortunately, particularly in countries characterized by significant racial and ethnic inequality, such genetic "predisposition" to crime is likely to be heavily racialized and gendered. In the United States, for example, which has the world's highest rate of incarceration, Black men are incarcerated at a rate almost seven times that of white men and almost three times that of Latinos. In Australia, between 2000 and 2008, incarceration of Indigenous people increased by 34.5 percent, an increase nearly seven times that of non-Indigenous people.

Several studies together have shown that increased rates of incarceration among people of color and Indigenous people has a great deal to do with racism throughout the criminal justice system, from arrest through sentencing (Alexander 2012). Sociologists have provided ample evidence that race impedes structural opportunities, such as schooling, long before somebody might engage in criminal behavior. What role might genetics play, if any, given these structural factors?

A CLOSER LOOK

Criminals in the womb?

In 2001, economists John J. Donohue III and Steven D. Levitt advanced what has come to be known as the Donohue-Levitt hypothesis: legalized abortion in the United States led to decreased crime rates. They claimed to provide evidence that up to 50 percent of reduced crime in the U.S. could be explained by legalized abortion.

While they argued that some of this reduction in crime could be attributed to smaller cohort sizes (i.e., fewer births and thus less crime), they also wrote that "teenagers, unmarried women, and the economically disadvantaged are also substantially more likely to seek abortions" (381). The implication of this claim is that the children of such women, had they not been aborted, would have been more likely to engage in crime.

This is a racialized and consequential claim. In the United States, African American women are disproportionately represented among poor and unmarried women. Unsurprisingly, the publication of the Donohue-Levitt hypothesis sparked controversy—on both methodological and cultural grounds—that remains unabated today, and many viewed the research itself as racist.

Of course, the suggestion that crime can be prevented by abortion raises the specter of genetic testing. What if embryos or fetuses could be tested for their future "propensity" to engage in crime, for example in the detection of a "crime gene"? Would this information actually benefit society? How might it harm individual women and their babies? Would it extend "racial profiling" to the womb?

Given the racialization of criminal justice practices, it is likely that women of color specifically would be targeted for genetic diagnosis for criminal predisposition in their fetuses. It is certainly *unlikely* we would see white women of the upper class targeted for genetic testing regarding the likelihood that, for example, their offspring would perpetrate massive financial fraud on millions of lower- and middle-class people.

Genetic testing for criminality offers a frightening combination of racialized theories about crime, such as the Donohue-Levitt hypothesis, and expanding reproductive technologies.

Clearly, cultural and legal understandings of genetics and criminal behavior are thoroughly grounded in social ideas about race, and in social structures and institutions such as the criminal justice system. Whereas a sociological perspective might look to structural factors that lead to excessive incarceration rates among black males in the United States and Indigenous people in Australia, such as inequality and lack of opportunities, genetic explanations assert that people are in prison because "nature" made them do it; they are inherently prone to criminality.

Sexual orientation

Akin to genetic foundations of "criminal" behavior are genetic explanations for sexual orientation and identity. *Homophobia* is the fear of and discrimination against people presumed to be homosexual. *Heterosexism* is the structural, cultural, and personal privileging of heterosexual identities and activities. Both homophobia and heterosexism, which may be expressed differently across different cultures but share common features, drive feelings and policies around human sexuality.

Seeing homosexuality as a genetic illness or medical condition has led individual scientists and biomedical researchers to seek a cause, and presumably a cure, for homosexuality. As historian Jennifer Terry (1999) argues, homosexuality has long served as an "abnormal" marker against which to define normality; in numerous studies, homosexuality was seen within medicine as a pathology or disease to be identified and fixed.

At the same time, some gay rights activists have also argued that sexual identities are ground in biology. They use these theories to prove they were "born this way" and therefore cannot be blamed for their natural inclinations. This is an important strategic argument in societies where there are strict regulations against homosexual behavior. Many legal systems are designed to protect the civil rights of people with immutable characteristics, such as race and gender.

Simultaneously, some anti-gay activists use the "born this way" framework to argue that lesbian, gay, and bisexual people are genetically flawed or biologically driven. This argument is also used in favor of a discriminatory agenda (Falomir-Pichastor and Hegarty 2013).

Research in the early 1990s by a team led by openly gay geneticist Dean Hamer (1994) claimed to find the so-called "gay gene" located on the X chromosome. Given eugenic histories, there may be reason to be concerned about such "definitive" proof of a gay gene, since subsequent decisions would likely be made about human genetic engineering or pregnancy termination.

Human geneticist Bryan Sykes (2003) has suggested that there is not a simple "gay gene" that exists in human populations, but rather there is a genetic association with homosexuality. He claims that a gay gene would be difficult to pass down from generation to generation since homosexual men would likely have fewer children than heterosexual men.

Across many political positions, whether or not there is a "gay gene" seems to be a preoccupation within contemporary cultural debates. Queer theorist Eve Sedgwick (1991) argued that the political and social urgency to eliminate homosexuals has influenced scientific research, advancing a biological model in hopes of prevention or extermination. Notably, genetic research on what makes people heterosexual—the presumed "norm" in most societies—does not capture the same level of scientific interest or media attention (Blank 2012).

As we shall see, interest in genetic causality extends to disease, including innovation of new diagnostic technologies.

Genetic testing for disease

Geneticists claim there are hundreds of diseases related to changes in the human genetic code; however, most of these diseases are extremely rare. Mutations in specific genes may prevent the genes from creating vital proteins or cause alterations in the proteins they produce.

Some examples of diseases that stem from genetic mutations include:

- cystic fibrosis
- Down Syndrome, also called trisomy 21
- Duchenne muscular dystrophy
- hemophilia A
- thalassemia
- sickle cell anemia
- polycystic kidney disease
- Tay-Sachs disease
- metabolic conditions such as phenylketonuria.

Genetic testing for disease, a form of surveillance, is significant because it is the evaluation of asymptomatic people (those without any symptoms) with the aim of improved health outcomes for the overall population. The ability to determine if genetic mutations may be present is provided by screening of infants, testing the potential parents (sometimes called carrier screening), and/or testing the embryo or fetus (Non-Invasive Prenatal Diagnosis or NIPD).

Currently, newborn screening is used widely for a variety of conditions although countries have divergent practices, with some testing for many diseases and others testing for few (Burke et al. 2011). While testing can provide information leading to preventive interventions, false positives and ambiguous test results may lead to decision-making based on potentially inaccurate information.

As genetic testing expands, with a corresponding rise in *genetic counseling* as an obligation for potential and expectant parents, it is likely that more pressure will be applied to parents, and to women in particular, to make the "right" decision. Depending on the results of NIPD, for example, a pregnancy might proceed with reassurance that the fetus is deemed "healthy." Or, a woman may be diagnosed as a high-risk pregnancy with a fetus that has tested positive for a genetic mutation. Will the pregnancy be terminated or medically managed?

As anthropologist Rayna Rapp (2000) and other scholars of reproduction have shown, a woman's ability to make decisions about prenatal testing and intervention varies with access to health insurance, availability of procedures, by race and class, geographically, and according to the level of social support they receive. At what point in the diagnostic process are women's own needs and desires overshadowed by state, legal, and eugenic concerns?

Similarly, disability studies scholars Erik Parens and Adrienne Asch (1999) explain that pressure to use prenatal genetic screening reinforces cultural preferences for certain types of bodies, leading to increased discrimination for people with disabilities. For example, a decline in the population of individuals with Down Syndrome is likely as testing and subsequent abortion (where available) becomes more prevalent (Benn and Chapman 2010).

Reproductive autonomy is challenged with NIPD because it creates new opportunities to evaluate fetal fitness. Categorizing fetuses as "defective" or "normal" has the potential to shape women's reproductive decisions (Casper 1998). Additionally, women are at greater risk of surveillance and coercion regarding their reproductive choices.

Feminist scholar Abby Lippman (1994: 29) has argued: "Prenatal testing is developed and applied with inherent expectations of how, when, and why it will be used that are tied to attitudes about women and about disability. It is currently expanding in a society that is deeply fractured along lines of gender, race, class, and ability. Its promotion and application can only reflect and, in turn, reinforce those divisions."

Babies "R" Us

Ideas about heredity have also created a near-universal social preference for genetic continuity in human reproduction, or the desire to reproduce children that are genetically related to the parents and especially to fathers. The growth of *assisted reproductive technologies (ARTs)*, spurred by technological innovations and decisions to delay child-bearing in the Global North, also reveals a preference for biologically related offspring.

Although adoption is a path to parenthood for many people around the world, those in the Global North who have access to boutique health care are more likely to pursue technologically-intensive pathways to parenthood. From *in vitro* fertilization (IVF) to surrogacy, a variety of technologies are being harnessed in the service of biological kinship, genetic stability, and reproduction (Mamo 2007).

ARTs have the potential to shake up traditional understandings of parenthood. In certain types of IVF, sperm may be combined with donor eggs in a laboratory and then implanted into a potential mother's uterus. Here, the resulting embryo would have a biological "father," a biological donor, and a gestational mother, and possibly two mothers if the receiving couple are lesbians. What would it mean to consider the donor in familial terms, especially given that she has provided half of the embryo's DNA?

In surrogacy, a woman may gestate and give birth to a baby that is not genetically related to her. United States and British courts have typically found that surrogate mothers have fewer rights in relation to the babies they grow and deliver as compared to the biological parents, illustrating the primacy of genetic connections in the legal system. Increasingly, women from the Global North are turning to Global South countries, such as India, for surrogates at less cost and with fewer regulations (Twine 2011).

The preference for our "own" babies, especially in the Global North, has multiple consequences for children and families. For one, a genetic preference may create tensions among siblings when some children are biological and others are adopted. For children who are adopted, the dominance of genetic explanations may induce a variety of feelings, including anxiety if they don't have ready access to their own biological information such as risk of disease.

Paternal preference for a biological connection means that, in the case of IVF, women may undergo expensive, invasive "treatment" to create a baby genetically related to the father, even if the baby has no genetic material from the mother. It is typically easier and cheaper for women to become pregnant from donated sperm than for an embryo transfer (using her own or others' donated gametes). What gender politics are operating here?

In many non-Western cultures, families may include both "immediate" and "extended" members in one group or clan. Here, biological connections are both more capacious and fluid. Genetic continuity may be understood less in terms of an individual or couple's preferences, as in the Global North, and more in terms of survival of an entire community or village.

In some cultures, however, where land rights, leadership, and other social benefits are passed down through generations, a biological connection is paramount. Yet these same practices may perpetuate gender and class inequalities, with sons favored over daughters and other entrenched practices. In fact, so important is eradicating these inequitable preferences, the United Nations has specifically operationalized gender equity as Goal 3 of the United Nations Millennium Development Goals.

Cross-cultural analysis helps us to understand that genetic meanings are not constant across time and space; nor do they universally assume the same social, legal, and biomedical importance.

Conclusion

Human genes are understood to be essential building blocks for bodies and increasingly for behaviors ranging from sex to racial difference to crime. Genetic frameworks guide our knowledge of how bodies work and what people do. Yet the expansion of genetic explanations for everything from physiological characteristics such as height to an individual's propensity to rob a bank has muted or wholly obscured social and environmental explanations.

If we understand all human beings to be shaped by genes, then there is little room for rearranging social structures or engaging in activism to expand opportunities for vulnerable members of society. This chapter has shown that our thinking may be so shaped by a biological metaphor, in this case "genetic," that we lose sight of other frameworks and paradigms for making sense of what it means to be human and what it means to intervene in human bodies and on the world around us.

See also

Cells and tissues; Digestion; Reproduction

Key terms

Human genetic engineering, genethics, biological essentialism, social constructionism, sociobiology, genetic determinism, genetic discrimination, eugenics, Social Darwinism, neoliberalism, race, social classification, racialization, genome variation, biological citizenship, genetic predisposition, homophobia, heterosexism, genetic counseling, assisted reproductive technologies, genetic modifications.

Discussion questions

1 Genetic explanations of human difference have dominated popular understandings of mental illness. Commonly referred to in terms of "vulnerability to mental illness" due to genetic components, scientists are still teasing out causal mechanisms. What are the social consequences if genes are found to play a

determining factor in mental illness and mental disorders? What type of interventions could and should be made available to people?

2 Human genetic engineering presents the possibility of manipulating human lives and futures, for both good and bad. What if scientists discovered genetic markers that determine skin tone? How might such a finding play into racial politics in your country? What are some ethical dilemmas connected to engineering for lighter or darker skin?
3 Asthma and allergies are conditions that have a strong link to genetic mutations. An employer is launching a new campaign to hire workers who can work in dusty and pollen-filled environments. The employer mandates that job candidates undergo a required cheek swab to rule out those with genetic predispositions to asthma and allergies. Is this a form of genetic discrimination? Is it justified?
4 What are some potential consequences of newborn screening? What are the costs and benefits?
5 Imagine you are 20 years in the future considering reproductive options. What do you think will be strongly recommended by the medical industrial complex prior to reproduction? What types of genetic testing do you predict will be available? How does the notion of the "good gene" seep into the potentially pregnant woman's body?

chapter 3

Cells and tissues

Introduction

Science fiction is full of stories about scientists harvesting biological material from people's bodies to grow new cell and tissue lines for use in wound healing, cloning, and for other purposes. And yet, these practices are not limited to fiction; from the development of cell culture techniques in the mid-20th century to the production and sale of cell lines such as HeLa (Landecker 2000), cells and tissues have traveled between bodies, laboratories, cultures, and nations.

As we learned in the last chapter, genes are widely considered to be the "building blocks" of human life, guiding everything from our appearance to behavior. However, biologically speaking, cells are, in fact, the real building blocks. They are among the tiniest life forms, and they serve as the architectural foundation for all organisms. Inside of each human being, for example, is approximately 10 trillion cells. Cells are made up of biological molecules (e.g., lipids, nucleic acids, proteins) and water, and they contain organelles and subcellular structures.

Our cells provide vital information that our bodies need to function.

In between the individual cells and the whole organism are other body parts, most notably for the purposes of this chapter, the tissues. A tissue is a collection of cells that share a similar function. Tissues that share identical functioning make up other body parts, such as organs. There are four basic types of tissues: epithelial tissue, connective tissue, muscle tissue, and nervous tissue. Each tissue type is responsible for a particular function or set of functions.

To put it more simply, cells form the base structure and are grouped into tissues of like functionality or purpose. Each tissue or set of tissues is distinct from other sets of tissues and their corresponding functions. None is privy to the workings of the whole organism. And yet, in intriguing ways, together their smooth functioning contributes to a healthy system (Gallucci and Matzinger 2001).

This structure, with its emphasis on functionality, has led to cellular structure being used as a model for social organization (Maasen, Mendelsohn, and Weingart 2001). From Auguste Comte to Herbert Spencer and Emile Durkheim to Talcott Parsons, "society" has often been represented as a functional whole made up of smaller units. Individuals, like cells, perform according to their required roles, thereby stabilizing the whole—social homeostasis. Individual bodies are likened to cells and combined they form the social body, which is akin to society.

Learning about cell morphology and function required certain technologies, specifically the microscope, and thus also required certain kinds of economic resources. After all, cells are not visible to the human eye without benefit of magnification. Dutch scientist Antonie van Leeuwenhoek is often credited as the "father of microbiology," recognized as having first "discovered" or viewed single-cell organisms (Bolfert 2007). He is also known for his discovery of sperm cells (Moore 2007).

Cells adapt, mutate, and reproduce, and they also may die. Some, such as neurons, live without dividing as long as the organism is alive; others, such as stem cells, never die but mutate into other kinds of cell. Without them, our bodies would be unable to do the things they need to, such as breathe and eat. The foundational status of cells in human physiology and conceptions of "life itself" is just one of the many things about cells that make them sociologically interesting. In the remainder of this chapter, we explore others, such as their use in metaphor, biobanking, medicine, immunity, and more.

A CLOSER LOOK

Post-9/11 and the uses of metaphor

Following the events of September 11, 2001, in New York City, in which a terrorist attack destroyed the World Trade Center and killed some 3000 people, a new term entered the American lexicon: terrorist cell. Gone are the days of the lone bomber or hijacker; today's terrorists are said to be organized, both locally and transnationally, into cells.

But what does this mean? Certainly, the media and governments are not referring to barred rooms, as in prisons and jails. In the context of terrorism, "cell" means something quite specific, and its use relies on a shared understanding of biology. As explained in a 2001 *Slate* article: "By dividing the greater organization into many multiperson groups and compartmentalizing information inside each cell as needed, the greater organization is more likely to survive if one of its components is compromised." Terror cells also invoke a creeping sense of doom; the metaphor denotes growth, expansion, and mutation.

How might we explain terrorism in the post-9/11 era without ready access to such biological metaphors as "cellular"? In a provocative article about global microstructures of terror, sociologist Karin Knorr Cetina analyzes various forms of organization including the cellular, the familial, and others. She is particularly interested in levels of complexity, and writes "that lightness of structure amounts to more than a lack of formal, rationalized institutional structures. It implies a transition to a temporal complexity of a kind where system stability depends on and arises from the fluidity and instability of component units" (2005: 226).

So just as genetic explanations and metaphors permeate our understanding of human culture and behavior, cellular metaphors give meaning to new forms of global violence. As in the previous chapter, we see how seamlessly "cultural" and "scientific" explanations can stand in for each other. The cell serves not only as a metaphor for organizational structure, but also as a model. As we noted in previous work on HIV/AIDS in Africa, there has been a conceptual shift from T-cells to terrorist cells in thinking about and responding to disease and contagion (Casper and Moore 2009).

Biobanking

Storing human cells and tissues in repositories is known as *biobanking* where biospecimens including blood, saliva, plasma, and purified DNA are maintained as libraries or archives of the human organism. Sharing data and materials means that scientists have access to more information than they would without biobanking or in their own individual laboratories. Collection and maintenance of multiple samples means more data to be analyzed; indeed, scientists have begun to speak of this as the era of Big Data. There are more than 300 million specimens in biobanks in the U.S. alone, with a growth rate of 20 million every year (Baker 2012). The rise of

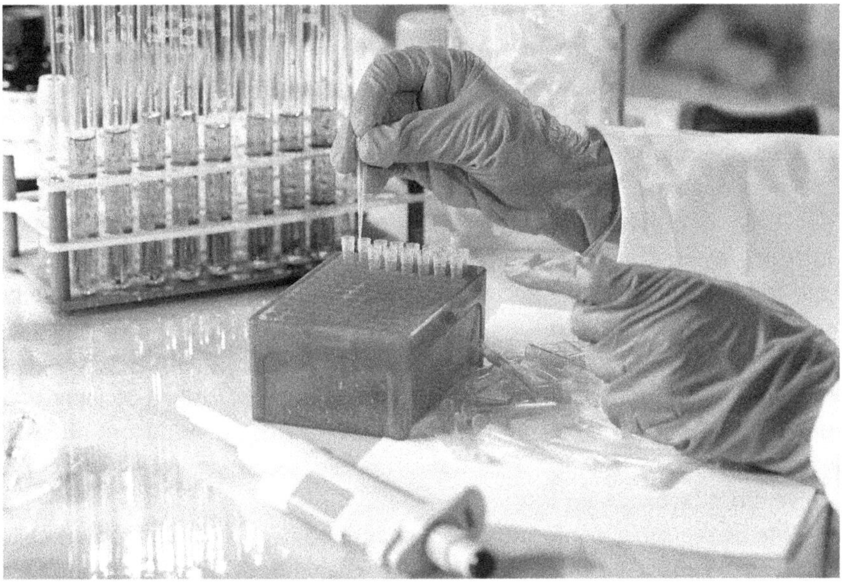

FIGURE 3.1 Drug testing
Image © Shutterstock

genomics (Fortun 2008) and personalized medicine (Hedgecoe 2004) has fostered this growth.

In 2011, U.S. science writer Rebecca Skloot published *The Immortal Life of Henrietta Lacks*, about the unknown woman behind the popular HeLa cell line, perhaps now the most famous example of biobanking. The book immediately became a bestseller and was translated into more than 15 languages including French, Spanish, and Portuguese. Skloot speaks widely in the U.S. and globally on the ethics of science and on the Lacks case, and the book generated significant debate and conversation (Nisbet and Fahy 2013).

Prior to publication of Skloot's book, few knew of Henrietta Lacks, an impoverished, African American tobacco farmer who died in 1951. Treated for cancer at Johns Hopkins in Baltimore, clinical staff retained her cells after her death; they were used to create the profitable HeLa cell line, which was used to develop vaccines, cancer medications, and more. The cells were used without her consent or knowledge, and as Skloot shows, Henrietta Lacks's surviving family (she had five children) did not know of their use until researchers began to study her descendants, too.

The tale of Henrietta Lacks, while certainly compelling, is not unique (Hoeyer 2008; Drabiak-Syed 2010). Numerous examples from around the world reveal that scientific work with human cells raises important questions about consent and autonomy, ownership of human body parts, and research ethics. Because we can harvest cells from people's bodies, does this mean we should? Regulatory bodies and ethicists struggle to answer this question (Allen et al. 2010), while markets proliferate. Cell lines, as use of HeLa has shown, are extremely important to scientific research on many fronts.

They are also extremely valuable. Although difficult to precisely quantify, it is estimated that HeLa alone generated hundreds of millions of dollars for companies, clinics, and others using the cell line (Truog, Kesselheim, and Joffe 2012). *Moore v. Regents of the University of California* was one of the earliest legal cases in the United States involving ethical and financial issues related to biobanking. Plaintiff John Moore, from whom tissue had been removed during leukemia treatment (without his explicit consent that it would be used for a cell line), argued that the line created from his cells was worth US$3 billion. The California Supreme Court ruled in the university's favor.

One of the most contentious issues in biobanking is that human cells and tissues must be removed from a person's body and processed in a laboratory. It may certainly be the case that an individual consents to their biological material being used in some settings. However, vast social inequality means that quite often, people with fewer resources, less cultural capital, and limited social power—such as Henrietta Lacks in the mid-20th century—find that their bodies have been mined in lucrative ways that fail to benefit them or their families.

This dynamic was on display in a widely discussed legal battle between Arizona State University and the Havasupai Tribe in Arizona, in which researchers used blood samples from the Havasupai to investigate issues beyond the original research protocol focused on diabetes. Ultimately, the university settled the case (Harmon 2010), de facto acknowledging that harm had been done.

Globally, biobanks that market "ethnic cell lines" pose additional concerns, especially for people of color and Indigenous people (Thacker 2005). Such practices have been termed biocolonialism, biopiracy, and bioprospecting (Hawthorne 2007), drawing on metaphors of mining and harvesting to critique global corporate science. In addition, in an era in which "race" is continually being reconstructed through genetic research (Roberts 2011), can we even speak of something as concrete as an "ethnic cell line"?

The proliferation of biobanking and related practices has prompted scholarly and popular attention to commercialization. Sociologist Catherine Waldby and cultural theorist Robert Mitchell describe these practices in terms of "tissue economies." They write: "At the macro-social level, tissues circulate through the social body according to technical principles that regulate their accumulation, transformation, multiplication, distribution, and utilization" (Waldby and Mitchell 2006: 33). Journalist Scott M. Carney has examined the transnational and largely unregulated "red market" in human body parts (Carney 2011).

Immunity and diseases

Cells and tissues play a significant role in the immune system (Schultz and Grieder 1987). Immunity is a complex process involving various structures in the body that protect against disease. The body has a layered defense system designed to seek out and eradicate any "foreigners," including viruses, bacteria, cancerous cells, and other pathogens. Leukocytes are white blood cells that act as cellular barriers to infection. One type, phagocytes, engulfs and kills intruders. Mast cells live in connective tissue, while neutrophils are found in the bloodstream. Immunity is often represented in terms of battle ("help your body fight off intruders"), while metaphors of immunity are used culturally (Martin 1995).

One of the justifications given for cellular and tissue research, including biobanking, is that new knowledge may help prevent or cure disease. Although cells throughout the human body are implicated in various health problems and many conditions can be understood at the "cellular level," a number of diseases directly involve or are caused by cells or changes in cells.

Cancer, for example, is typically caused by abnormal cell growth. Cells may mutate, or divide and grow in an unregulated manner. This cell activity can produce cancerous tissue, tumors, and other malignancies. The specific etiology of cancer is complex, and there are many types of cancers that affect different body parts. Cancer is understood to be due to environmental, genetic, and epigenetic factors (Chik, Szyf, and Rabbani 2011).

Some, such as breast and ovarian cancer, show an increased risk in family members, suggesting a genetic basis (Bennett 1999). Others are considered preventable; for example, lung cancer due to air pollution and smoking (Danaei and others 2005). *Epigenetics* refers to alterations in the structure of chromosomes that does not relate to mutations in the DNA itself. A good example of this is reversible modifications of histones, the proteins around which DNA is wrapped in eukaryotic cells, which can lead to significant alterations in gene expression.

For more than half a century, there has been a "war on cancer," a dedicated and heavily supported biomedical initiative to eradicate cancer. However, these efforts—transnational in scope and funding—have failed to locate a cure (Davis 2007). As a number of scholars have argued, the expensive biomedical war on cancer has detracted attention from prevention efforts, such as cleaning up the environment (Leys 2009).

Some cancers have become intensely politicized. For example, breast cancer emerged in the 20th century Global North as one of the most feared, discussed, and funded women's health issues (Lerner 2003). Critique from women's health activists centered on biomedical attention to genetic factors over environmental causes, as well as the emergence of a "cancer industrial complex" (King 2008).

Particular critical attention was paid to dubious partnerships between industry and social movements (Klawiter 2008). Taken together, all of these critiques called for a feminist, woman-centered, preventive approach to breast cancer rather than a continued biomedicalized search for the cure.

And yet, while a cure for cancer remains elusive, it is also deeply desired by many. The World Health Organization reports that 7.6 million people died from cancer in 2008, that 70 percent of cancer deaths occur in low- and middle-income countries, and that 30 percent of cancer deaths are preventable. Cancer accounts for 13 percent of all deaths globally, with lung, breast, colorectal, stomach, and prostate cancers responsible for the majority of deaths.

As with other health issues, morbidity and mortality are stratified. Although cancer epidemiology is complex (Jemal and others 2011), in general poor people around the world with less access to preventive screening and treatment have a higher mortality rate. In the UK, for example, "cancer rates are higher for deprived groups" (Gordon-Dseagu 2006). In a number of countries, cervical cancer—a leading cause of death for women—incidence reveals striking inequalities in screening and care (Todorova and others 2009; Gakidou, Nordhagen, and Obermeyer 2008).

Mitochondrial disease is caused by failures in the mitochondria of cells. With the exception of red blood cells, all of the body's cells contain mitochondria, specialized structures that generate the cell's energy supply (Henze, Martin, and Martin 2003). The mitochondria are responsible for converting the vast majority of chemical energy from sugar into energy that is immediately useable for cellular function. These structures generate approximately 90 percent of the energy needed to sustain life. When mitochondria fail, less energy is produced within the cell, leading to cell injury and death.

Most of the body's parts and functions may be affected by mitochondrial disease including the brain, nerves, muscles, kidney, heart, liver, ears and eyes, pancreas, and more. When more than three organs or systems are involved, mitochondrial disease is suspected. Both children and adults can be stricken with mitochondrial disease, but diagnosis is difficult, requiring specialized laboratories that may not be available in some parts of the world (Chinnery and Turnbull 2001).

Researchers estimate that mitochondrial disease is far more common than previously understood (Schaefer and others 2004). However, the epidemiology of mitochondrial disease is challenging, in part because patients affected by mitochondrial disorders may have symptoms resembling those of other conditions (e.g., diabetes mellitus).

A CLOSER LOOK

Sickle cell anemia

Sickle cell anemia is a disorder that causes red blood cells to become sickle-shaped, like a crescent (Bunn 1997). Usually red blood cells are round like a donut and they carry hemoglobin through the body, delivering oxygen from the lungs. In sickle cell anemia, the hemoglobin is abnormal, causing the crescent shape of the cells. These cells are sticky and do not flow through the body as easily as normal red blood cells.

Normal red blood cells live for about 120 days in the bloodstream. Sickle cells live only 10 to 20 days. They cannot be replaced quickly, and so the body is perpetually short on red blood cells, producing anemia and a variety of health problems.

People with sickle cell anemia can become dizzy, short of breath, experience headaches, feel coldness in their extremities, and may become jaundiced. Episodes of pain may occur when the sickle cells block blood flow to limbs and organs. These are called crises.

More serious problems include infection, splenic crisis (too many red blood cells in the spleen), acute chest syndrome (similar to pneumonia), pulmonary hypertension (elevated blood pressure), stroke, eye problems, gallstones, leg ulcers, and multiple organ failure (U.S. National Heart, Lung, and Blood Institute). While there is no cure for sickle cell anemia, there are treatments including antibiotics, fluids, oxygen therapy, and even blood transfusions—where these treatments are available.

Globally, millions of people are affected by sickle cell anemia. Because it is a genetic disease, the offspring of people with sickle cell may also be affected. The World Health Organization reports that the people most affected have ancestors from sub-Saharan Africa, South America, Cuba, Central America, Saudi Arabia, India, and the Mediterranean countries.

In the United States, about 72,000 people are affected, predominantly African Americans. The incidence rate is about 1 in 500 African American births, according to the World Health Organization.

Studies have shown that carriers of one sickle cell anemia allele are resistant to malaria infection (Aidoo et al. 2002). These findings offer an evolutionary rationale for why the sickle cell alleles are maintained in a population. Malaria is caused by a parasite, Plasmodium, which is transmitted to humans through mosquito bites. Untreated, it can be deadly. The World Health Organization reports that, in 2012, malaria resulted in approximately 627,000 deaths, primarily among African children.

Sickle cell anemia can be diagnosed with a blood test. Although newborn screening is recommended, access to this technology is stratified by class, race, and geography. Public health scholars have called for screening to be part of basic health services, but this is not always possible in rural, poor areas (Modell and Darlison 2008).

In addition, early screening in the United States in the mid-20th century led to concerns among African Americans about racism and eugenics (Atkin and Ahmad 1998). These concerns prompted the Black Panther Party, a radical activist group, to offer its own health services, including screening. As sociologist Alondra Nelson (2011: 118) argues: "The health education dispensed by the Black Panther Party contextualized sickle cell anemia within a matrix of mediating factors that included not only biology but also racism and poverty."

Thus, sickle cell anemia reveals social inequalities, global health disparities, and the ways that bodies, health, and illness can become springboards for activism and social change.

Politics of stem cells

Stem cells are the biological material from which other, specialized cells are generated, and they can be harvested from bone marrow, fat, and other bodily sites. Human embryonic stem cells (hESC) are harvested from fetal tissue, typically only a few days old. Stem cells may be useful in preventing and treating diseases, such as Parkinson's and diabetes, and healing injuries such as spinal cord damage. Embryonic stem cells are especially appealing because they can grow and adapt to new environments.

However, not all stem cells are created equally; or rather, not all stem cells are *interpreted* equally. Because some cells, the hESC, are taken from human embryos, ethical and political questions have emerged. Moral and political concerns hinge on the status of human embryos, as either persons or biological research materials (Casper 1994). Thus, human embryonic stem cells have been especially controversial and subject to regulation (ISSCR 2006).

One contentious issue is the relationship of pregnancy termination to stem cell research, especially where abortion is highly politicized. In other words, where do the embryos come from? Scientists and health advocates in Europe called on the European Parliament not to succumb to anti-abortion factions and eliminate funding for human embryo research (Boseley 2012). In the United States, conservative "pro-life" groups have vocally opposed embryonic stem cell research, suggesting that the moral cost of harvesting embryos after abortion is not worth potential gains in human health (Reaves 2001).

One "work-around" for harvesting embryos is to use the embryos that are "left over" during *in vitro* fertilization. During IVF, often several embryos are created, but only a few are implanted in a woman's uterus. The remaining embryos may be stored, donated, or discarded. These embryos may be used in stem cell research. Indeed, use of "leftover" embryos is seen by many (but not all) ethicists as a compromise position that accords moral respect for the embryo while also facilitating scientific research that might save lives (de Wert and Mummery 2003).

Another issue that has provoked ethical concerns is *stem cell tourism*, or traveling to a different country to seek stem cell therapy (Lindvall and Hyun 2009). As with medical tourism overall (Bookman and Bookman 2007), stem cell tourism often involves lower costs and quicker, easier access to therapies. Destination countries for U.S. "tourists" include those with less regulation, such as Mexico, Peru, Russia, and China (Sesin 2013). Often, treatments offered in these countries have not been approved by the U.S. Food and Drug Administration, which makes stem cell tourism especially appealing to people from the United States.

Different countries, reflecting diverse political frameworks and meanings of embryos and fetuses (Morgan 2009), have enacted various kinds of regulations to address controversy while also supporting scientific research (Walters 2004). Within the European Union, consensus has been difficult given differences across nations, with the UK adopting a liberal policy and some countries (e.g., Bulgaria, Ireland, Germany, Italy) banning hESC research entirely. Israel is a world leader in hESC research, while in Asia, China has one of the most liberal policies. Globally, hESC regulation is a patchwork, with no standard policy (ICSCN 2008).

Beyond the issue of types of stem cells is the related issue of how stem cells will be used. Cloning, especially of organisms, is deeply contested and regulated. There are several types of cloning: molecular cloning, or modification of DNA molecules; cellular cloning, or proliferation and/or modification of cells in culture, such as HeLa); and organismal or reproductive cloning, or creating a new organism of identical genetic structure to the original.

Ever since the cloning of Dolly the sheep in 1996 at the University of Edinburgh, the practice of organismal cloning has especially captured the interest of scientists and the human imagination (Franklin 2007). Because reproductive cloning presents the possibility of creating genetically identical organisms, it both fascinates and troubles—and it strikes at the heart of Western philosophical notions of individuality.

Cloning also disrupts fundamental tenets of many religious beliefs (Bainbridge 2003), including the notion of a soul and the belief that the universe and all life within it were created by a deity. For example, would a human clone have a soul, or would it be a soulless version of you, as the question was framed by *Scientific American* (Bering 2009). As a number of scholars have suggested, debates about human cloning reveal deeper tensions between science and religion (Bainbridge 2003; Shepherd et al. others 2007).

Conclusion

As this chapter suggests, in many contexts—scientific and otherwise—cells are considered the building blocks of human life. And yet, they represent so much more. For certain foundational social scientists, cells are metaphorically productive and useful for examining and explaining human relationships and structures. Our linguistic use of the term "cell" is used expressively to describe a coherent group operating in concert, much as the body's cells operate to form a coherent whole.

We also explored here how the transnational *biomedical industrial complex*—the aggregate of biomedical enterprises, suppliers and individuals that comprise the "health" industry and value capital accumulation—since late capitalism, contests the ownership of cells harvested from individuals' bodies and finds them potentially highly profitable. Once out of the body, who owns cells and tissues—the originator or the extractor? The individual or the corporation? In the case of sickle cell anemia, genetic explanations cleave with cellular understandings in explaining this blood-based disorder, while also—as the Black Panthers illustrated—attending to racism and its embodied effects.

Finally, as technologies are developed to harness the potentialities of stem cells, including hESC, we expand our capacity for regeneration of human bodies and organs, and we make possible such practices as cloning. As we have shown, these cellular technologies and "tissue economies" bring with them a host of political and bioethical questions.

See also

Genes; Organs; Fluids; Bones and muscles; Central nervous system

Key terms

Biobanking; epigenetics; embryo; stem cell tourism; biomedical industrial complex

Discussion questions

1 Metaphors of biology, such as "terrorist cell," are widely used. Can you think of some other, non-scientific ways in which the term "cell" is used in your social context? What are the various meanings?
2 How would you react to finding out that your biological material had been stored and used following a hospital procedure without your explicit consent? Why would you react in this particular way?
3 The history of screening for sickle cell anemia provoked concerns about scientific racial profiling and eugenics. Can you think of any diseases now that might provoke similar concerns if mass screening were in place?
4 The term "savior sibling" has been coined to describe the situation of parents conceiving a child in order to use their biological material in pursuit of a cure or treatment for an already existing child. If you could become pregnant (or impregnate your partner) for the sole purpose of creating a savior sibling to harvest their fetal tissue or other biological materials (i.e., bone marrow) for use to save a genetic relative, would you do so? Why or why not? What are the major ethical issues here?
5 If you could clone yourself, would you do so? Why or why not? What about cloning individual organs? What might be the negative consequences of such an action?

chapter 4

Organs

Introduction

Consider how you will never see most of your organs and, unless something goes wrong, you rarely consider their existence but rather operate on faith that they are inside of you and behaving correctly. In fact, most of us will live our entire lives never having seen most of our organs.

Organs are typically defined as a differentiated group of tissues working together to perform a specific bodily function. Some organs act individually, while others function together as essential, interrelated parts of bodily systems, such as digestion and circulation.

The American Association of Anatomists reports there are about 7,500 individual named human body parts, and that more than 1000 of these could be classified as organs based on a broad definition. More specificity in what organs do, particularly as part of systems, results in a number closer to about 60 individual organs or pairs of organs.

Some commonly recognized organs include the heart, brain, spleen, stomach, liver, pancreas, small and large intestine, skin, uterus, kidneys, bladder, and eye. While human beings can live without some of these (e.g., spleen, kidney, eye), most are essential to healthy functioning of the body and the sustenance of life.

Elsewhere in the book we highlight a few individual organs, such as the brain, skin, genitals, heart and lungs, and the spinal cord. This chapter, while examining the role of several major organs, also examines meanings and uses of organs in human cultures. What, we ask, do we make of organs? How do they figure in social life? We consider the role of transplantation medicine and the existence of "red markets" in organ trafficking. Also, the very notion of "the end of life" is radically altered by the technological imperative of organ replacement.

But where better to start than with the organs we eat?

A CLOSER LOOK

Let them eat cake

For both of us, being children meant savoring favorite "special" dishes. Liver and onions, a delicacy shared with a beloved Danish grandfather, or Fegato alla Veneziana (calves liver Venetian style) prepared by an Italian grandmother. The liver—and of course, it was beef liver and not human—was smothered in onions and fried or sautéed, resulting in a succulent, tender slab of meat that could easily be sliced with the curve of a fork.

From these beginnings of growing up consuming liver, as well as the juicy innards of the Thanksgiving turkey once it was cored out before being stuffed, we both became vegetarians for over twenty years and then pescatarians (fish eaters), and we've only recently reintroduced some limited meat into our diets. The mere thought of consuming offal is revolting to these former liver-eaters, suggesting that what and how we eat has shifted substantially across our life-course.

Humans in a variety of cultures have long enjoyed eating the insides of animals, such as livers, stomachs, brains, testicles, lungs, kidneys, and hearts. Organs are a staple protein source in many parts of the world. Recently, these items have been championed in the Global North as highly beneficial to athletes and others interested in "healthy" eating. For example, the so-called Paleo Diet recommends organs because they are full of proteins and vitamins, as do many fitness magazines aimed at people trying to build muscle.

While some humans are seemingly happy to chomp down on the organs of nonhuman animals—even the brains of monkeys, who are among our closest genetic kin—most of us would recoil at the idea of consuming human organs (or other body parts). And yet, like many species of nonhumans (Kristal et al. 2012), humans have sometimes engaged in placentophagy, or the eating of the placenta after birth. (Other methods of handling the placenta include burial, drying, and incineration.)

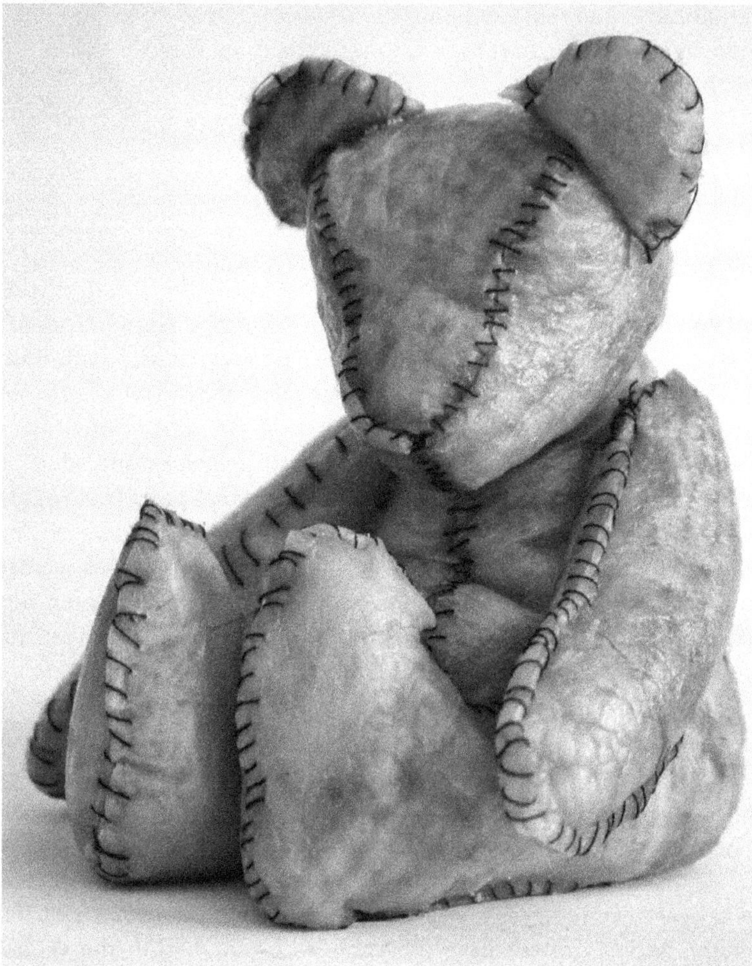

FIGURE 4.1 Placenta teddy bear
Twin Teddy by Alex Green, image © [re]design

One of the most fascinating things about the placenta is that it is not a permanent organ. The placenta develops only during pregnancy, after the blastocyst implants into the maternal endometrium. The placenta continues to grow, with maternal blood supply to the developing fetus complete by the end of the first trimester. Through the umbilical cord, it allows for circulation of blood and nutrients to the fetus, and also enables processing of waste.

The placenta, sometimes called the "afterbirth," is expelled from a pregnant woman's body during childbirth, although sometimes it must be manually removed. By no means a widespread practice, eating the placenta is believed to reduce postpartum depression and other complications of childbirth, and also may fulfill certain cultural rituals of birth. In one study, women who ate their placenta reported increased mood, energy, and lactation (Selander et al. 2013).

With access to the Internet, one can easily find a variety of placenta recipes at mothering, health, and "alternative" living sites. The word "placenta," it should be noted, derives from the Latin word for cake.

Classifying organs

As Linnaeus and others have shown, humans tend to classify "nature" into categories based on visual and other markers. These classifications, as we discuss in other chapters, carry with them a range of cultural meanings and histories. However, as science and technology studies scholars Geoffrey C. Bowker and Susan Leigh Star (2000) argued, classificatory systems—like standards—also flatten and obscure differences and variation.

Organs are typically classified into vital, semi-vital, non-vital, and vestigial organs. *Vital organs* refer to those without which we would die: brain, heart, lungs, liver, at least some kidney. Vital, etymologically, derives from the Latin *vita*, or life. Thus, vital means life, and the term "vital" is used not only to describe organs that sustain us, but also life itself.

Non-vital organs are those we can ostensibly live without: ovaries, uterus, testes, gallbladder, and so on. However, even so, their loss might require substitutions of that which they provided. For example, the ovaries and testes are sites of hormone production in the human body; people who lose these organs may, for example, need to undergo hormone replacement therapy.

The category of semi-vital organs is sometimes used to describe those we could live without for some period of time, but that we really need in order to thrive. For example, these organs include the intestines, spleen, and skin, among others. Regarding skin, people with third-degree burns, in which most of their skin is damaged, die at much higher rates than those with less severe burns (Pavoni et al. 2010). Over 90 percent of all burns occur in the Global South, and children typically have higher mortality rates than adults (Erickson et al. 1991).

So while anatomists and transplant specialists speak of vital and non-vital organs, it would seem that losing some parts of our bodies not deemed "vital" could pose major problems. Indeed, who gets to decide what body parts are "vital"? We might survive without some body parts or damage to others, but our functionality would be limited. How, then, do functionality and "quality of life" enter into conversations about vitality? As sociologist Heather Laine Talley (2014) shows, our understandings of "vital" and "non-vital" inform how we think about transplants, including their promotion as life-saving technologies.

The category of vestigial organs deepens the complexity. These are organs that are believed to have lost their functionality and importance due to evolution. One rationale for deeming these organs as vestigial is the "proof" that a human can live without the organ. The appendix, the spleen, the tonsils and adenoids have all been deemed "junk parts." However, some scientists believe the bio-scientific community has yet to develop research tools or methods to identify the function of these "irrelevant" parts.

Transplant medicine

Sometimes, our organs give out. We may experience organ failure due to disease, age, trauma, wear-and-tear, or other circumstances. If the failing organ is non-vital, then we're typically not facing a life-threatening situation. However, if vital organs fail, our demise may be imminent. Without functional organs, technology, such as dialysis machines used when kidney function is compromised, can keep us alive for only so long before we die. Thus, survival under such circumstances is contingent on the availability of technology and ultimately, of replacement organs.

Surgeon and medical historian David Hamilton (2012: 1) offers a rich, historical account of transplantation, noting that replacement of body parts appears in ancient mythology "from Iceland to Africa." As do hybrids, or what he calls "chimeric monsters." Think here of the centaur, sphinx, Ganesha, and Pegasus. He suggests these early fantasies of creatures that were half-human and half-beast presaged the development of the field of transplantation, beginning in earnest in the 18th century.

While a full history of transplant medicine is well beyond the scope of this chapter, it is worth noting that new technologies of healing and repair were consistent with Enlightenment ideologies of human advancement. The Scientific Revolution is recognized as the emergence of "modern" science, and medicine is certainly encompassed in this recognition. Although high-tech "modern medicine" is typically dated to the 19th century, the Enlightenment ushered in the dawn of *scientific* medicine.

Transplant medicine, historically as now, relied on the availability of replacement body parts, whether from a patient's own body (such as a skin graft) or from the body of another. Most often, organs and other body parts were procured from newly dead donors. Although the first skin transplant was performed in 1869, and the first cornea transplant in 1906, it was not until 1968 that the Uniform Anatomical Gift Act was passed in the United

States, providing a legal mechanism for a living person to arrange donation of organs after his or her death.

The ability to prevent immune rejection was a major boost to transplant medicine. Prior to development of pharmaceuticals, such as cyclosporine, which suppress the immune system response, the recipient's body might reject the new organ as a foreign body. This is the immune system at work. Transplant rejection can lead to organ failure immediately or over the long term; if the reaction is acute and immediate, the patient might die if the organ is not removed. Even with immunosuppressants, recipients may still "reject" their new organs.

Considering the long, rich history of procuring and transplanting human body parts, the practice remains deeply entangled in cultural beliefs, ethics, economic issues, transnational inequalities, local and global regulations, and an all-too-human "visceral" reaction to using the body parts of another human being. Despite the proliferation of transplants and the success of transplant medicine, there is still an "ick" factor.

Replacing human body parts continues to evoke, for many people, a sense of fear about the loss of individual identity and about science-run-amok (Morgan et al. 2008). Such fears were fantastically expressed in Mary Shelley's 1818 novel *Frankenstein*. And yet transplantation, once the domain of science fiction and horror, is now *de rigueur*, and carries with it a set of ethical and political issues that bear consideration.

The traffic in organs

Knowledge of organ transplantation is fairly widespread. Most people, especially in the Global North, seem to know that a patient suffering organ failure can have a healthy organ transplanted into their body—most commonly a kidney, lung, or liver. The details vary by country, with different nations regulating organ donation and transplantation in different ways.

What people seem to consider less often, and what the media seem to report less often, are the circumstances under which organs are provided for patients who need them. How, in other words, are organs procured? From whom do clinicians get organs? Are they paid for or donated? Who regulates the traffic in organs?

Legal organ procurement is typically regarded as altruistic, or done for unselfish reasons and not compensated (Delmonico et al. 2002). A living donor may provide an organ for a family member, for example a sister who donates one of her kidneys to an ailing sibling. Most often, however, organs of various types are procured from dead or living strangers and regulated through governmental practices, such as waiting lists and donation networks.

In Australia, for example, the DonateLife Network is national, but operates in state jurisdictions. In the U.S., the federal Organ Procurement and Transplantation Network maintains the only waiting list in the country. Other countries have similar national networks, such as Give Life in Canada and the Japan Organ Transplant Network. These networks and/or agencies maintain records, keep track of available organs, and monitor clinical practices.

Globally, only one nation (Iran) allows organs to be transported beyond national borders; in all other countries, organ procurement happens locally, within the country itself, and is not a commercial enterprise. But there is considerable global interest in organs, given issues of illegal trade and medical tourism. The World Health Organization, for example, maintains the Global Knowledge Base on Transplantation.

The flourishing underground market in human organs, which operates "off the grid" of regulation, also operates well beyond standard ethical practices—and national borders. Although data collection is difficult given that the trade is illegal, there is some information available, including major recipient and donor countries (Shimazono 2007). India, Pakistan, and China are among the major exporters of organs, while major importing countries include Australia, Canada, Israel, Japan, Oman, Saudi Arabia, and the United States.

Tracking illegal organs means following the money, as organs have become embedded in transnational commodity chains. From the "donor's" body—often a poor person in need of money for survival—to the recipient's body, one can track the movement of organs through markets, brokers, hustlers, transit networks, and shady clinical operations. Anthropologist Monir Moniruzzaman (2012) refers to the illicit organ trade as a kind of *bioviolence*, in which poor organ sellers are often tricked and suffer immensely.

End-of-life decision-making

Organ failure can be a precursor to death, and thus transplantation invokes important questions about decision-making at the ends of life. That we can replace failing organs with healthy ones from donors does not necessarily mean that we should. After all, transplants are expensive, technologically intensive, and available only to some people in the world, those with the most resources. Economic calculators from 2011 in the United States range from US$200,000 to US$1.2 million for transplants not inclusive of lost wages, or travel and lodging to and from transplant locations.

Sometimes organs fail due to age; they simply wear out after being used throughout a person's lifetime. With a limited supply of organs, how should decisions be made about the rationing of healthy and available organs?

Should an elderly patient suffering from organ failure be denied a transplant simply because that person is likely to die sooner than a younger patient? Should organ transplants be reserved for the "healthy" and young, or should they be allocated fairly, by immediate need rather than age or health status? Who should make these decisions? Doctors and lawyers? Patients and their families? Legislators?

And what of patients who need organ transplants because of their own behavior, for example a heavy drinker whose liver fails, or a longtime smoker with damaged lungs? Should they be "rewarded" for their supposedly deviant behavior with a chance to start over? What if lungs are damaged by industrial pollution, such as in coal mining and production of chemicals? Should the industries themselves foot the bill for health consequences of occupations that can lead to increased morbidity and mortality?

Such "end-of-life" questions are typically the domain of *bioethics*, a field of inquiry and practice that reflects on and attempts to resolve health care dilemmas. Bioethics emerged from some of the most horrific practices of the 20th century, particularly Nazi medical experimentation on human beings (Proctor 1990). The Nuremberg Code was developed in response to these abuses and identifies basic principles of medical research including informed consent and the duty to avoid suffering (Annas and Grodin 1995).

However, even with these principles in place, ethical decision-making is challenging, especially at the beginnings and ends of life where uncertainty prevails, and especially in relation to new technologies. Often, technologies make possible scenarios that have not yet been ethically or socially vetted. An excellent example of this is human cloning; the technology to clone living creatures proceeded well in advance of our collective ability to make sense of it (Nussbaum and Sunstein 1999).

An especially important issue related to transplant medicine is what constitutes death. If an organ fails, is the person dead? What if the failing organ is the brain? Anthropologist Margaret Lock (2001) argues that the emergence of the category of *brain death* helped to facilitate the expansion of transplant medicine. After all, if donor bodies are technically dead because their brains are defined as dead, then their organs can be "ethically" harvested.

In the United States, the category of brain death has been used since 1968. As Lock notes, in Japan it has only been in use since 1997, and then under limited circumstances. Transplant medicine has been slow to grow in Japan, in part because of the restricted conditions under which the category of brain death can apply. Lock's work, like that of other anthropologists who study aging and death (e.g., Cohen 1998), suggests that death is always more than simply biological, and that social and cultural meanings are important.

In recent years, bioethics has been challenged to move beyond the laboratory and the clinic, and into the broader social, political, and economic domains in which decision-making actually happens (Weisz 1991). An exemplary project in this regard is Rayna Rapp's (2000) study of women's decision-making about amniocentesis. Here, race, class, gender, and health care access are all part of the decisions women make.

Currently, as we discuss next with respect to transplant tourism, bioethics is as much about geopolitics, transnational commodity exchanges, and global inequality as it is about what happens in clinics.

A CLOSER LOOK

Medical tourism

Although historically some people have traveled far from home to seek medical treatment, such as elites who departed urban areas for sanatoria or curative "rest homes" in the countryside, *medical tourism* across national borders has proliferated in recent years, becoming a US$60 billion industry (Evans 2008). Patients from wealthier countries such as the United States and Canada travel to countries in the Global South, such as Thailand and India, seeking more affordable care.

Patients travel for both surgery and long-term care, often paying considerably less than they would at home. For example, a heart bypass that costs $144,000 in the U.S. costs $26,000 in Costa Rica and $10,000 in India; a hip replacement that costs $43,000 in the U.S. costs $14,000 in Mexico and $10,000 in India (Pilgrimed.com). Even after factoring in the costs of travel, this is an enormous saving.

Little wonder, then, that the medical tourism industry is booming, with its own travel agents, magazines, professional associations, and websites. However, critics have argued that while the savings are immense for patients from Global North countries, the costs for people in the destination countries may be too high. For example, countries that invest in technology and personnel to serve medical tourists may do so at the expense of providing basic health care services for their own citizens (de Arrellano 2007).

Transplants are increasingly among the procedures obtained through medical tourism, posing unique risks and ethical considerations. Although donors in destination countries are sometimes already deceased, often the donors are living. Thus, in transplant tourism wealthier (although not necessarily wealthy) patients travel to other parts of the world to obtain organs from poor people (Scheper-Hughes 2011), both reflecting and sustaining the commodification of body parts and unregulated global economies.

When patients travel to a country such as Mexico or India for transplants, there is no guarantee that the organs they receive have been obtained ethically (Cohen 2013). In all but one country (Iran), it is illegal to sell organs across national borders; yet, within countries there are vastly different regulations governing how organs are obtained and from whom, and in many there are robust "red markets" (Carney 2011). China has faced considerable international censure for its use of organs removed from executed prisoners (Danovitch and Shapiro 2011).

Transplant tourism thus fuels organ trafficking and deepens existing global inequalities (Budiani-Saberi and Delmonico 2008). In addition, there may be clinical risks including inadequate postoperative care, insufficient follow-up, infection, immunosuppression, and transplant rejection. Patients traveling between countries are at greater risk for multi-resistant bacterial infection (Rogers et al. 2011).

Organs by other means

Because there are more people who need fresh organs to survive than there are organs available for transplant, and also because of the ethical issues involved in organ procurement and transplantation, scientists and clinicians are looking to artificial and regenerative organs.

Artificial organs are manufactured devices made of materials such as plastic, titanium, and other substances that are designed to replace "natural" organs. Like prosthetic limbs, artificial or prosthetic organs are placed into the body and ideally function as the original did. Some examples of artificial organs include the cochlear implant, eyes, the heart, and the trachea.

Recently, scientists have developed *bioartificial* (also called biomechanical) organs from stem cells. These organs are made of synthetic material that has been "seeded" with human cells (Fountain 2012). Using a patient's own cells to create the bioartificial organ minimizes organ rejection, as the patient's immune system doesn't recognize the organ as an intruder. A synthetic trachea was successfully transplanted in 2004 (Macchiarini et al. 2008).

Regeneration of organs is also a possibility. All species are capable of renewing and regenerating, although to varying degrees. For example, while humans slough off and grow new skin and hair, starfish (a type of echinoderm) can generate new arms, geckos can make new tails, and the planarian flatworm can split into two distinct clones of itself. Such regenerative qualities are exciting with respect to transplant medicine.

Scientists researching the topic suggest that mechanisms of regeneration, once understood fully, can pave the way for therapeutic interventions. J.A. Baddour et al. (2012), for example, investigate a number of human body parts capable of renewal including the thymus, adrenal gland, thyroid gland, intestine, heart, lungs, blood vessels, liver, germ cells, nervous system, eye tissues, hair cells, kidney and bladder, skin, hair follicles, pancreas, bone, and cartilage. They explore the regenerative capacities of each part, as well as the likelihood that a body part will be amenable to bioartificial intervention.

Sociologist Nikolas Rose (2012), among others, has asserted that biology is no longer destiny. Perhaps nowhere is this more obvious than in research on *organ regeneration*. If, indeed, humans—like starfish and planeria—are someday capable of producing new organs from old and continually regenerating, then the very terms of human existence, including mortality, may shift. What would it mean to our understandings of death to renew our body parts, and thus to extend our lives? Clearly, we would need to consider who has access to these extended lives and at what costs to others.

Conclusion

Unlike many other chapters in this book, considering organs may invoke a certain level of revulsion. From eating the placenta to harvesting organs from warm corpses, the mind wanders to dark images from science fiction or biomedical laboratories. Generally not given a second thought, unlike our hair, faces, or even genitalia, we take for granted the everyday functioning of our organs.

However, organs reveal themselves when not functioning optimally or failing completely, and humans have used a combination of biomedical innovation, biotechnology, and personal sacrifice to establish procedures for transplantation. Deepening existing inequities, legal and illicit organ procurement and medical tourism illustrate how the have-nots are physically in the service of the haves.

Death has taken on new meanings, too, with the possibility of providing replacement organs for the privileged few. What it means to "take care of oneself" might have a different valence if one is a potential donor or future recipient. The nascent field of organ regeneration, while possibly eliminating the fears of harvesting organs from the poor, brings a host of new bioethical considerations.

See also

Cells and tissues; Bones and muscles; Central nervous system; Circulation; Respiration; Digestion; Reproduction; Skin; Genitals

Key terms

Vital organs, transplant medicine, bioviolence, bioethics, brain death, medical tourism, organ regeneration, bioartificial

Discussion questions

1. What are your thoughts about the placenta as an organ? Would you ever consider eating the placenta following childbirth? Does eating a placenta bring to mind the taboo of cannibalism? Why or why not?
2. Under what circumstances would you consider being an organ donor? If you feel strongly in opposition, is there anything that could incentivize your decision? Do you think people should be paid for their organs? Or should organs only be given away for altruistic reasons, such as to a family member?
3. Have you ever heard any urban legends about organ thievery? If so, what were they? What do you think they represent?
4. The illegal organ trade exists because some people are willing to pay for healthy organs. Is there a way to intervene at the "demand" level such that the organ "supply" might be curtailed? Alternatively, should economic resources in biotechnology be dedicated to cloning technologies as a way to generate organs? What are some social implications of such biotech infusions of monies for cloning?
5. Who is worthy of an organ transplant? Do people who engage in so-called deviant behaviors, such as drinking alcohol or smoking cigarettes, deserve to have their failing organs replaced with healthy organs? Why or why not?

chapter 5

Fluids

Introduction

This chapter explores diverse meanings of the term *fluid* in biological, sociological, and medical settings. Following a brief exploration of what fluids are, we examine the status of those who manage body fluids and we discuss the ways that exposure to body fluids can taint individuals. Adopting a wider angle of vision, we look at the social management of bodily fluids through the lens of the sociology of waste.

Next, building on the ways bodily fluids change individuals' social status, we examine cases of injury or disease where the control of body fluids is made possible, such as through a stoma. We then explain the ways the transmission of body fluids connects humans through illness. Finally, we conclude with a discussion of the commercial exchange of body fluids and, in particular, the buying and selling of human semen.

Experiments with water

Depending on where you grew up, you may remember those primary school science lessons about the states of matter. You were instructed to take careful notes during lectures on the properties of the three classical states: solids, liquids, and gases. Then, in the lab portion of the lesson, water—typically the most accessible substance for experimentation—was presented in a glass beaker. You recorded observations of the liquid as it swished around, spilled, dripped, made things wet (including your clothes). Placed inside a freezer, the liquid became ice, a solid. Boiled on a Bunsen burner, possibly the most fun activity for budding young scientists, steam indicated how the liquid turned to gas. We loved learning that these states of matter—liquid, gas, and solid—all exist within human bodies, too.

In this chapter, we are especially focused on liquids and gases. The ability to flow, to take the shape of a container, to maintain a certain density—these are material qualities that categorize substances as liquids or gases. Liquids and gases are also fluids. In fact, the human body is 60 percent fluid—or water, mostly from *intercellular fluid*. An inventory of bodily fluids might include, for example, tears, blood, saliva, sweat, mucous, lymph, gastric acid, cerebrospinal fluid, vaginal secretions, urine, semen, female ejaculate, breast milk, amniotic fluid, feces, and vomit. These individual body fluids are deeply interconnected to the social body and its "waste."

Reading this list may evoke feelings of disgust and revulsion—but let's consider this for a moment. What is it about your own (or others') bodily secretions and lubricants that sickens? All the other members of our species have fluids and must attend to them, and yet fluids feel so private and potentially humiliating should they become public. The intimate and carnal qualities of bodily fluids, their odors and appearance, the emotions they convey, and the vulnerabilities they reveal all influence our reactions to our own gases and liquids.

Varied meanings of fluid

Fluid is also a term we use to describe social situations. When "fluid" is used to describe a situation or a thing, we mean to emphasize that the situation or thing is apt to change at a moment's notice; the current state of affairs is not stable or fixed. Fluid situations can bring about individual and cultural anxiety since uncertainty implies a loss of control. As a loved one lays unconscious in an ICU, having a health care provider tell you that the current medical situation is "fluid" is probably not very reassuring. Ironically, however, in social life most things *are* fluid and changeable, as we live in a world of many moving parts with ongoing emergence and flux.

The concept of fluidity is also meaningful to social scientists and cultural theorists as global travel, neoliberal markets, and social networking almost daily alter our experiences. Fluidity implies there is an easy flow between ideas and a sense of movement in and across positions without strict adherence to a static location or framework. Sociologist Zygmut Bauman (2005: 2) examined the notion of *liquid modernity* whereby "liquid life is a precarious life, under the conditions of constant uncertainty." In other words, because of the speed at which capital enterprises attempt to extract profit, there is a quickening of the rise and fall of pursuits, consumer goods, and social trends. As a result, individuals must become flexible and act as if there is limited reliability and rigidity in social institutions. We live in an increasingly fragmented world of quick-paced production and social media networking—so we must adapt to this liquidity and adjust to its uncertainty.

Sometimes, the term *sexual fluidity* is used to indicate how some individuals' sexual orientation and sexual behavior might change over time and within different social situations (Diamond 2009). For example, prisons are sex-segregated institutions; this means that while incarcerated, individuals who do not identify as "gay" or "lesbian" might engage in homosexual relationships. When released, a person could return to heterosexual partners. Over the course of a person's life, sexual orientation can be fluid.

In sum, the term "fluid" is culturally and linguistically useful. As a noun, body *fluids* are material productions of the human body rife with meaning and purpose. As an adjective, *fluid* is used to describe social situations, and as a theoretical concept *fluidity* implies nonfixity or movement through different identities and ideas. These ideas of fluidity are useful for social scientists because they interrupt the rigid bounded categories of identity and imply there is a porous and diffuse flow as individuals and interactions change over time and by region.

As examined in myriad ways throughout this text, the human body, its needs and behaviors, in many ways serves as a template for how we organize our everyday lives. Here, we focus on four different theoretical and empirical contributions to the sociology of body fluids: social management of fluids, their control, their transmission, and their commercial use.

Social management of body fluids

Scholars are particularly interested in relationships between symbolism and the social and personal management of body fluids. To this end, the germinal work of anthropologist Mary Douglas contends that "the body provides the basic scheme for all symbolism" (1966: 163). These symbols located in and on the body represent socio-cultural markers that both situate people and guide how we behave with one another. Meanings attributed to bodily fluids provide clues for engaging in social interactions. These collective representations are transmitted in part through medical-industrial complexes (e.g., meanings of disease) and health care encounters (both *allopathic* and homeopathic).

Sociologist Norbert Elias' historical interpretation of manners and etiquette identifies the processes that facilitated the emergence of the modern self within a civilized (controlled) body and how this development relates to state formation. Elias demonstrates that different modes of behavior such as bodily carriage, bodily fluid management, and table manners change as part of an ongoing interactive process between individuals and larger structural formation. His work adroitly illustrates the mind/body/society relationship from the Renaissance onward as "civilizational self-controls" became internalized, causing people to notice not only themselves, but also "the other" and to adjust their behaviors.

Civilité is dependent upon seeing, and being seen by others and behaving properly in social situations. The civilizing process reveals that socially acceptable behavior is connected to social structure and the emergence of a self-controlled individual. Manners and bodily etiquette originated from the upper social strata, eventually filtering down to daily interactions of people from all social classes. Elias maintains that social control was "mild" in courtly medieval society as compared with later eras. For example, courtly table etiquette dictated that people could spit while eating (being sure to refrain from spitting across the table), and eat from others' plates as long as one refrained from "falling on the dish like a pig, and from dripping bitten food into the communal space" (Elias 1994: 87).

Centuries later, table manner etiquette changed rather significantly, as social controls became more exacting for the individual within public space. By 1859, spitting was considered to be completely unacceptable. Not only did people become detached from one another into individual social spaces vis-à-vis separate bowls, some people became "so delicate that they (did) not wish to eat from a dish which others have dipped a spoon already used" (Elias 1994: 87). Saliva now had different social meanings and meant that social space became differently regulated. The civilized body is physically separate from others ("my" space), self-reflective, controlled, and aware of how it must behave in any given social context—in both public and private realms.

A great deal of everyday life is taken up with managing our own or others' bodily fluids. Through disciplinary power heavily employed in childhood about the self-regulation of body fluids, we become part of civil society. From going to the bathroom, to changing a tampon, to blowing your nose, to grooming and deodorizing, to quenching thirst, to wiping away tears, the micro-management of bodily flows makes up part of each day. (See, for example, an anthropologist's reappropriation of garbage and waste in living environments http://cityoflivinggarbage.blogspot.com/.)

Much of this management of bodily fluids is privately accomplished by individual people performing self-care, or by others (especially women) caring for small children or dependent adults, such as those with certain disabilities. However, flushing a toilet or vomiting in a classroom, for example, are activities in which personal management of fluids gives way to public coordination through odors, sounds, or other signals.

Waste water

Simultaneous to our seemingly contained and private bodily management, there is also social, political, and environmental management of body fluids.

For example, in industrialized parts of the world, waste treatment plants collect water from municipal lines and "balance" the bacteria, chemicals, particles, and other pollutants. Through a process of oxygenation, purified water is then released back into the environment. Unless our toilet is stopped up or the city sewer line breaks, we give little thought to the journey of body fluids once they have evacuated our bodies and we have flushed.

Fluids quickly travel from the microsociological to the macrosociological, from the individual body to the global body. We are composed of fluids, require them to live, expel them from our bodies, and manage them in micro-interactions and institutional structures. Our fluids are part of the global ecology. And yet, in many parts of the world, people do not have access to flushable waste receptacles.

Below is a photograph taken from a wastewater treatment plant where miles of sewage lines bring dirty water and run it through a series of pumps

FIGURE 5.1 Sewage plant
© Shutterstock

and filters for purification. The remaining sludge is spun at high speeds to remove inorganic matter, and the solid organic waste is pumped through bacterial decomposition to create bio-solids that may be used as fertilizers on farmland.

Global flows of waste

The *sociology of waste* is an emergent sub-specialty in development studies. It tracks the transnational flows of waste—not exclusively embodied fluid waste, but waste, nonetheless—that is produced by human life. As sociologist Honor Fagan (2002) states: "Waste can be conceptualised as a globally circulating fluid, its production and management governed well beyond the nation state."

The trend in waste management policy is for greenhouse gases, ozone depleting gases, and toxic waste *to flow from* the Global North to the Global South. Since industrialization, the toxic load of handling global waste fluids—sludge, sewage, trash, chemical byproducts—has increasingly fallen to those geographical regions with the least global power. So routine was the shipment of hazardous waste to "less developed" countries, the United Nations amended the 1989 Basel Convention, an international treaty to enforce nations to manage their own waste.

Moving back to the level of social interaction, fluid management, or the ways in which we create and marshal larger institutional resources to control body fluids, entails cooperation among public agencies, health care practitioners, federal regulators, skilled laborers, and most of all, low-paid semi-skilled or unskilled workers. Individuals who experience the most frequent and proximate exposure to body fluids often rank lowest on scales of social stratification.

Cleaning body fluids

For sociologists, one of the most interesting aspects of managing bodily fluids is how they link up with the concept of *dirty work*. Writing in 1951, sociologist Everett Hughes coined the term to refer to jobs or tasks that are generally regarded as degrading, disgusting, or vile. Scouring bedpans, changing and disposing of diapers, cleaning bathrooms, removing hospital waste—these tasks are not part of aspirational job descriptions nor are they typically key to anyone's "get rich quick" schemes.

Rather, these tasks are often seen as a dimension of the invisible labor that has been synonymous with women's work or the work of people of color. This gendered and racialized division of labor leaves the tasks of bodily management to the private, domestic, and uncompensated (or lowly compensated) sphere of primarily women and girls. In many settings across North America and Europe, from hospitals to nursing homes to domestic labor, this sphere is inequitably racialized, with poor women of color at the bottom rung.

When bodily tasks are unable to be performed by the individual and are made relational or public, other people are employed to manage the body fluids of those typically of higher status. Sociologists Shirley Drew, Melanie Mills, and Bob Gassaway (2007) investigate how those who engage in "dirty work," in particular the management of bodily fluids such as feces, bile, vomit, and blood, are liable to be tainted. They are tainted not just by the physical performance of cleaning and sanitizing spaces after these fluids spill (and thus potentially at risk for contagion), but they are also tainted by the social relationship of servility that is required of lower-status occupations.

Sociological analysis of the domestic and global management of body fluids thus reveals the larger workings of social, economic, and cultural power including occupational and family hierarchies. That is, the burden of managing waste falls on the least developed countries, just as dirty work is performed by and taints lower-status workers.

Controlling body fluids

As humans, we sweat, cry, sniffle, leak, ejaculate, pee, defecate, and bleed. For anyone who has ever had to potty-train a reluctant and potentially tantruming toddler, the littlest humans are essentially unpredictable geysers. Slowly over time, through processes of socialization, we transmit social norms and customs about body fluids to members of our group. And groups do have different norms and customs. We are often socialized to avoid taboo subjects and activities (taboos are the creation of prohibitions of particular activities, activities that bring such revulsion that they are forbidden), such as playing with feces or drinking human urine.

A great deal of growing up and becoming an adult is establishing normative routines about self-care—toileting, hygiene, eating, drinking, and sexual expression. In many ways, becoming an adult and a socially normative member of a community is training the body not to leak, or at the very least being skilled with prophylactics—diapers, panty liners, breast milk pads, bandages, condoms, antidiarrheal medicines, tissues, antiperspirants, and antacids (where these are available). Indeed, a thriving and lucrative industry anticipates and prevents body fluids from being seen or smelled by others.

Bodies leak. They create and disrupt meanings through the leakage. Leaky bodies disrupt the socially expected bounded rationality of many cultural norms. Feminist philosopher Margrit Shildrick (1997) finds that female bodies, in particular, are constructed within biomedical and bioethical discourses as leaky; the female body is deemed messy, unbounded, shifting, porous, flexible, and unpredictable.

Anthropological fieldwork demonstrates cross-cultural experiences of women's bodies as polluting because of their leakiness—"a bleeding vagina and dripping nipples testif[y] to its inability to remain in control" (Tsoffar 2004: 10; e.g., Farmer 1988). Shildrick and Price (1999: 3) argue that this leakiness contaminates rational and bounded spaces: "As the devalued processes of reproduction make clear, the body has a propensity to leak, to overflow the proper distinctions between self and other, to contaminate and engulf."

Although this leakiness has traditionally been depicted in pejorative terms, Shildrick's work attempts to resurrect leakiness as an asset in times of poststructural subjecthood, belonging to a subject that is nomadic rather than stable. Feminist theories about leakiness can be a springboard for understanding other bodily fluids, such as semen, urine, and breast milk, as soiling social spaces (Moore 2007b).

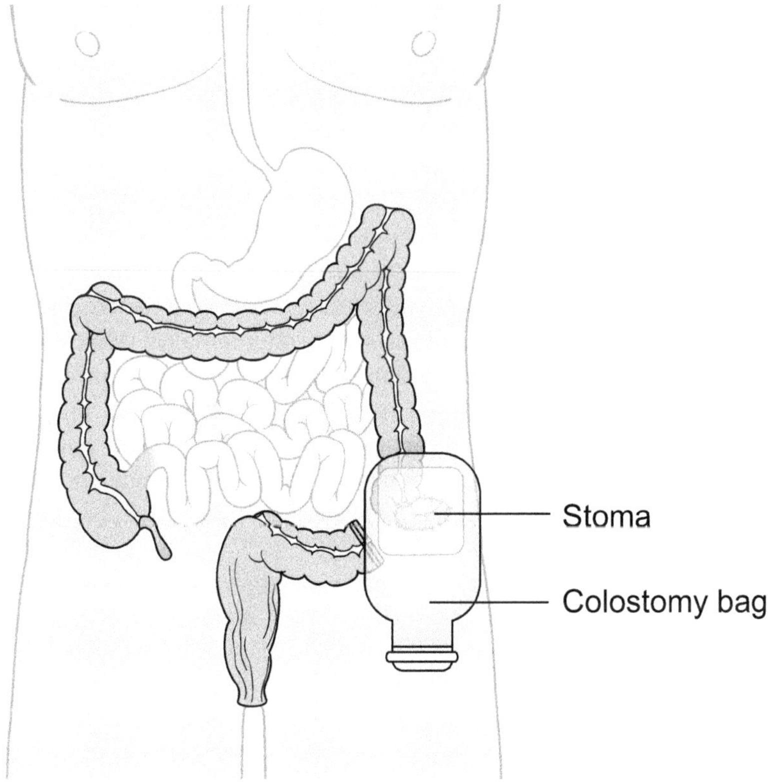

FIGURE 5.2 Colostomy pouch
Photo courtesy of Cancer Research UK and taken from the patient information website of Cancer Research UK: http://www.cancerresearchuk.org/cancerhelp

A CLOSER LOOK

Stoma and fluid bypasses

When the body does not manage fluids as it is expected to, medical intervention may be required. A *stoma*, or a surgically constructed opening of the body, bypasses internal organs. Some examples of stomas include colostomies, ileostomies, ureterostomies, or tracheostomies.

In the case of stomas used as alternative systems for fluid removal, the individual may have an attached bag that collects urine or feces. According to medical anthropologist Lenore Manderson (2005), one in 1000 individuals in the developed world has a stoma. Her research interprets how individuals with no bowel or bladder control remain sexual after stoma surgery. Her use of the concept of a "boundary breach" is used to explain how disruption of the taken-for-granted containment of body fluids also disrupts everyday personal relationships.

Of the 32 Australian men and women Manderson interviewed, all felt their lives were improved by addressing incontinence through stoma surgery. However, the stoma and the bag that collects internal fluids created tension for her informants, particularly in intimate, sexual situations. Many feared that the bags would leak and smell or make rustling sounds.

As one informant explained about the differences in her life: "Basic things like, if you have sex, is there a risk of the stoma coming off, I mean the bag coming off, or will it hurt you? I mean internally, by losing your rectum, are you the same person internally? What effect will it have to have a penis pressing against the walls of your vagina when there is no rectum? And if you have got that vigilance there as well, and this affects sexual response . . . you are lying on the bed and half of you is on auto-pilot and the other half of you is thinking, hang on, is the bag still safe, or, oh yuk, I can hear it go gurgle, gurgle or squish, squish."

Being a "properly" socialized human means the ability to normatively control and contain the flows of bodily fluids. A stoma interrupts and transgresses this social expectation and requires a renegotiation of personal self-concept, intimate relationships, and embodied expressions of intimacy. Other studies have indicated that over time (more than one year), however, those with stomas report improved self-esteem and an increasing degree of sexual confidence (Gloeckner 1984).

Transmission of body fluids

Humans have a very long history of spreading diseases through transmission of viruses and bacteria in bodily fluids. Today, we think of the exchange of body fluids as having to do primarily with sexual interactions, and in particular HIV/AIDS. However, human urine and feces mixed with shared water sources, as well as mucus projected through coughing and sneezing, can lead to the transmission of typhoid, hepatitis A, amoebic dysentery, rotavirus, giardiasis, cholera, salmonellosis, and shigellosis.

One of the greatest modern achievements of public health has been the creation and use of sanitation—leading to the immediate decline of transmissible diseases, both viral and bacterial. In the early 20th century in the United States, economists Cutler and Miller (2005) explain, clean water technologies, specifically filtration and chlorination, were responsible for dramatic reductions in mortality.

However, as the authors also note, these innovations in public health are not universally available across the globe: "Worldwide, roughly 1.1 billion people lack access to clean water and about 2.4 billion people do not have adequate sanitation. Cutting the share of people without suitable water and sanitation in half by 2015 is one of the ambitious objectives set by the Millennium Development Goals" (Cutler and Miller 2005: 21).

Body fluids are also transmitted during sexual activities, especially intercourse. HIV, hepatitis B, hepatitis C, gonorrhea, syphilis, chlamydia, and HPV are among the types of illnesses that are transmissible. A quarter century ago, HIV/AIDS was considered a major public health crisis and discussed widely in the Global North. It was a frightening epidemic that, after initial inaction, generated changes in activism, public health practices, clinical research, funding streams, and sexual politics (Patton 1990; Treichler 1999). Yet, when the disease was brought "under control" in the U.S. (a myth), for example, media and policy-makers stopped paying attention to HIV/AIDS.

Based on demographic analysis, the epidemiology of HIV/AIDS in the United States follows different trajectories when considering social class, gender, race, and access. For example, in 2011, young African American men (13–24 years old) had the greatest burden of disease and were disproportionately affected (Centers for Disease Control 2013). Two decades ago, there was at least one HIV/AIDS story daily in major newspapers and often more, whereas now the disease is rarely discussed in U.S. mainstream media.

Given that HIV/AIDS is more complicated and dangerous than it ever was, affecting millions of people around the globe and especially in the Global South, lack of media attention by Global North countries demands analysis.

Comparing fluids—water with semen, for example, or water with blood—helps to reveal the different meanings attached to liquid substances. Clean drinking water is a major public health goal, and yet corporations may benefit, in the absence of regulation, from being able to pollute the water supply. For some groups, clean water may not be a priority, and other concerns—such as profit—come to the fore. Sexual fluids are often considered contaminated, not simply because they can carry toxins, but because sex itself is perceived (although not universally) as dirty. In some settings, blood is seen as life-saving, whereas in others, it is considered to be tainted and thus subject to strategies of avoidance. Global diseases such as HIV/AIDS and environmental disasters may shape overall how cultures view and handle fluids.

In short, transmission of body fluids helps to make visible the varied meanings of sexuality, relations between people and nations, cultural definitions of infection, and shifting understandings of disease. To focus on the movement of fluids between and among people is also, then, to focus on human relationships and social structures.

Commercial use of body fluids

In the 2011 documentary, *What's My Body Worth*, British film-maker Storm Theunissen explored legal and ethical issues that stem from selling one's own body. Her research suggested that in the commercial and pharmaceutical marketplace, body fluids (e.g., urine, mucus, saliva, breast milk, blood, semen, ear wax, blister fluid, tears) sold from a living donor every month could yield over US$8000.

The application of commercial body fluids varies depending on the type of fluid. For example, urine is sold as "clean" for individuals who may use drugs but need to pass a drug test. Human saliva is used by pharmaceutical companies for use in drug development and experimentation. Blood has many uses, most notably for transfusion.

There is also an altruistic form of human fluid banking exemplified by the donation of excess breast milk. Human milk banks (HMB), as explored by Australian scholar Katherine Carroll (2012), collect expressed breast milk from lactating women and distribute the milk, once pasteurized, to infants, especially in neonatal intensive care units.

Milk donors are highly screened for contagious diseases as well as "lifestyle factors" to ensure the quality of their milk (Virtual Medical Center 2012). This "liquid gold," as Carroll calls it, is a valuable fluid that sustains an infant's life and, as our next section explores, can also ensure the health of the social body.

A CLOSER LOOK

Sperm banks and social marketing

Over the past century, one of the most fascinating commercial uses of human body fluid has been semen. The historical emergence of human semen cryopreservation (freezing semen) dates back to Italy in 1776. In the United States, the first documented semen bank began in 1950 at the University of Iowa. These early banks were exclusively part of university systems until the early 1970s. The university provided the materials, funding, and legitimacy to further explore, research, and develop clinical applications, leading to the first work on humans in 1954.

In 1972, commercial semen banks were established for infertile men and heterosexual couples. This new wave of semen banking converged with the modern birth control movement, which included an option for men, vasectomy. Men did not, however, respond enthusiastically to the push for vasectomies and the growth of semen banks did not meet expectations.

Some men, especially those vasectomized or undergoing chemotherapy, today use semen banks as a storage facility for their own future use—similar to a safety deposit box for precious valuables at a financial bank. This type of "fertility insurance" is marketed to alleviate fears of exposure to toxic substances harmful or lethal to sperm and/or men. Although presumably not the intended audience, single women and couples (lesbian and heterosexual) also use semen banks in order to reproduce.

Moore's (2007a) research on the sperm banking industry led her to coin the term *technosemen* to define the latest new and improved bodily fluid that semen banks advertise to clients which is purportedly better than naturally produced and procured semen. These are technical, medical, and mystical processes that are carefully presented to potential clients and described in great depth.

For instance, semen analysis includes sperm counts, morphology, motility testing, functional testing, and sperm washing ranging from the swim-up methods to percoll or retrograde. Each test is accompanied by a price tag. In addition to semen analysis, which purports to be a method of insuring "fertability" of the product, semen undergoes disease and genetic testing to diminish possible fears of genetic or contagious conditions, such as HIV/AIDS. These practices of semen analysis and disease testing create technosemen.

Marketing technosemen can challenge the assumed inelasticity of the semen trade by convincing, or better yet guaranteeing, the general public that technosemen is fertile, uncontaminated, and genetically "engineered" for desirable traits. In the age of epidemics, cloning, and increased environmental disasters, semen banks can capitalize on their low-tech and relatively low-cost procedures and so increase their revenues.

Thus, semen banks participate in, reinforce, and bolster the public's fear about material threats to the human race. Both semen banks and consumers would like to believe in the fantastic potential of semen. "People come to us because they want the best possible sperm for their future offspring," said one semen bank administrator. "We don't offer the village idiot, because we don't get much demand for such."

The processes by which technosemen is produced and advertised to consumers are deeply connected with how the "contents" of semen are stylized. Donor catalogues organized by race and ethnicity help to guide women's choices regarding reproduction. Rather than using a turkey baster or having sex with a partner, women are encouraged to strip down in sterile white rooms, lay on a table, put their feet in stirrups, and allow a physician to place frozen pellets as close to the cervix as possible.

As Lisa Jean explored with Marianna Grady (Moore and Grady 2014), the U.S. sperm banking industry is at a unique crossroads of social marketing in the 21st century. As use and diversity of fertility services grow, sperm banks are able to offer their customers more services than ever before. Likewise, the development of Web 2.0 provided sperm banks with distinctive pathways to new and current consumers. This confluence of biotechnological innovation and web-based social media is not exclusive to the sperm banking industry; these shifts affect all biomedicalized industries in the early 21st century.

The growth of social media outlets has spurred development of fresh, competitive strategies available to businesses, a development that has influenced the practice of several cryobanks. As a result, the use of Facebook, Twitter, and You Tube, for example, has begun to reorient social norms surrounding procurement of semen. It is not surprising to see a Facebook page designed by a semen bank with status updates about the availability of a popular donor's inventory. These status updates are punctuated with "likes" and hopeful commentary about future offspring.

Biomonitoring

Body fluids have economic and social value. One obvious example is the creation of reproductive marketplaces (e.g., the trade in semen). Perhaps less obvious is the use of human fluids for surveillance of human populations. *Biomonitoring* involves the measurement of trace compounds in human beings using molecular biological techniques. It is an extractive technology; that is, liquids and tissues are removed from human bodies so they can be manipulated and assessed. Whole bodies are not measured, but rather their leaky and diaphanous bits and parts, such as breast milk and blood, are measured. These fragments stand in for and represent whole persons, and indeed entire communities and the species writ large (Casper and Moore 2009).

Biomonitoring refers to "analytical measurement of biomarkers in specified units of tissues or body products . . . These biomarkers are any substances, structures, or processes so measured that indicate an exposure or susceptibility or that predict the incidence or outcome of disease" (Albertini and others 2006). Blood, urine, breast milk, semen, and other tissues and fluids are studied for the presence of cellular changes.

Biomarkers of exposure involve detection of original chemical contamination, while *biomarkers of effect* involve measurement of biological responses. Exposures are complex and may be synergistic; that is, chemicals together may produce adverse effects.

Breast milk is an especially useful fluid for measuring toxins in the body. Toxins accumulate in our bodies by attaching themselves to fatty tissues, which in women's bodies can be the same tissues that hold nutrients found in breast milk. Women are considered to be at greater risk from toxins given the higher fat content of their bodies vis-à-vis those of men (Vesely 2003), with breasts being a principal site of accumulation. Also, breast milk is one of the few "exit routes" for chemicals and may contain higher concentrations of persistent toxic chemicals than other fluids or tissues (Steingraber 2001). The presence of chemicals in breast milk is troubling because of the consequences for women, in terms of risk for breast cancer and other diseases, but also for infants who receive the off-loaded toxins with their mothers' milk.

In the 1970s, biomonitoring was part of a campaign to prevent childhood lead poisoning. When scientists and policy-makers learned that high levels of lead in blood could result in cognitive deficits, they advocated for removal of lead from gasoline, paint, and other products. The lead-free movement was successful in the U.S., although children in poor urban neighborhoods are often still exposed to lead in old paint, and leaded gasoline is still used throughout the Global South. Lead also persists in soil worldwide and may pose serious health risks (Schettler et al. 1999).

Biomonitoring produces a measurement referred to as *body burden*, or the total amount of chemicals present in the human body at the time of testing. There are numerous ways toxins enter the body including ingestion, inhalation, and absorption. Pregnant women can pass substances to their fetuses through the placenta and umbilical cord. Some chemicals stay in the body for only a short time, while continuous chemical exposure can create a "persistent" body burden by lodging in fat tissue, semen, muscle, bone, brain tissue, and other organs. DDT, a chlorinated pesticide, can remain in the body for up to 50 years.

Biomonitoring can reveal an individual's unique "chemical load," highlighting the kinds of chemicals we are exposed to every day.

Conclusion

Tracking the emotional, social, biological, economic, and geopolitical flows of human body fluids is integral to a comprehensive understanding of social life. As individuals we are trained practically from infancy in social norms about the maintenance and control of our body fluids. Indeed, a leaked fluid can stigmatize an individual with consequences beyond mere humiliation to social ostracism or strict sanctions—as will be seen in Chapter 12, in our discussion of obstetric fistula.

Although the human body is mostly comprised of fluid, especially water, the global supply of clean drinking water remains inequitably distributed and profoundly at risk of diminishing. Does privatizing water secure *all* people's health and wellbeing? How might understanding the movement of human fluids, both in and out of our bodies, contribute to our knowledge of social organization and geopolitics?

We started this chapter by noting the abundance of water in the human body. However, human bodies can and do soil the water, making it, in some instances, toxic to human bodies. Technical intervention is required to "treat" water so as to be clean again and potable, at least for some consumers. Globally, clean water is in short supply, in part because of "natural" factors such as climate change and in part because of human actions from pollution to privatization.

How ironic that something so fundamental to our lives—without which we could not survive—is so scarce for so many.

See also

Cells and tissues; Circulation; Digestion; Reproduction

Key terms

Fluid, liquid modernity, sexual fluidity, allopathic, sociology of waste, dirty work, stoma, biomonitoring, body burden, technosemen

Discussion questions

1 Why do you think there is such tremendous social stigma regarding human body fluids? Do certain body fluids elicit greater social revulsion than others, and why are some bodily fluids more taboo than others? Offer a sociological explanation for the disgust that surrounds some body fluids.
2 Thinking about "dirty work" and body fluids, what are some commonalities among people who are responsible for handling, cleaning up, and disposing of body fluids? How does direct management and disposal of bodily waste "soil" the relative social status of the individual?
3 Loss of control over one's body fluids means that individual bodies can seep into social life, often at unpredictable times. What are all the ways we manage our body fluids to try and avoid this occurrence? For example, make a list of all the products that one can use to contain their bodily fluids. How can our attempts to control our body fluids be derailed? How does fluid leakage also relate to social stratification and access to products to control seepage of bodily fluids?
4 One of the fascinating things about body fluids is the fact that they can continue to exist outside of the body, leaving a trace of the actual source or whole body. What are some examples of how residual body fluids might link an individual to larger social meanings? How do body fluids assume "lives of their own" and, at the same time, implicate the individual body from which they came?
5 Biomonitoring may be socially valuable and necessary, but it may also be coercive. Can you suggest a program of biomonitoring that would make the practice more just and respectful of the individual bodies that provide fluid?

chapter 6

Bones and muscles

Introduction

In human anatomy, bones and muscles are the structural foundations of the body. They form our shape, strength, and body type, and together they help to engineer movement and force of locomotion. In sociology, *structure* is a concept that refers to organized social arrangements that create patterns for social engagement. Structure, both anatomically and sociologically, refers to relatively obdurate realities that affect what is possible, and what is not. It is the structure that gives rise to the composition of the human body and to the order of social relations, although in both cases there is a range of variation depending on differences among individual people and the amount of agency they can wield in the world.

Cultural meanings such as worth, value, perseverance, loss, and virtue, are ascribed to our muscles and bones, often in highly gendered and racialized ways. For example, large biceps on men are seen as masculine and evidence of working hard. Bones and muscles thus reflect the gendered structure of many human societies as depicted through physical performance and competition of male and female bodies in homosocial spaces (e.g., athletics). Our bones and muscles are called on by our embodied experiences to function with a quest toward self-optimization within contexts of global capitalism. We examine these meanings of embodied social life as we delve into the physiological scaffolding of the human form.

In this chapter, we first define and describe human bones including medical conditions such as osteoporosis. We next move to an analysis of how bones become evidence of large-scale human conflicts as well as sources of popular entertainment and social justice movements. Jumping to muscles, we look to some popular cultural examples of muscles, and then move to a discussion of the sociology of sport and gendered/racialized notions of athletic achievement including the marketing and use of muscle enhancements. The chapter concludes with an explanation of social and anatomical tendons that link structures to movement.

The thigh bone . . .

Among the five- to eight-year-old set in certain cultural contexts, *Dem Bones*, a traditional spiritual song written by James Weldon Johnson and adopted from a biblical passage in which Ezekiel visits the Valley of the Dry Bones, is much-loved and very catchy. *The heel bone connected to the foot bone, the foot bone connected to the leg bone, the leg bone connected to the knee bone, and so on.* Singing the song with accompanying hand motions has the power to focus and unite (i.e., to discipline) an unruly crowd of children into a well behaved group through anatomical choreography. While the song does not catalogue all 206 bones in the human body—although some kids (and adults) have likely tried to do so—it does effectively introduce children to the skeletal architecture inside their bodies.

As we grow through our teens and into adulthood, we sometimes encounter an offering of culturally specific compliments: he has "chiseled" good looks or she has "great bone structure." Or the ultimate compliment, uttered in the romantic/sexual contexts of some youth: "I want to jump his (or her) bones." What are bones, though, especially in this context? It might be surprising to learn that bones are living and growing tissue comprised of collagen, which is the protein framework, and the mineral calcium phosphate. This is why, in the Global North at least, we are urged (often by the dairy industry) from a fairly early age to drink our milk and take our daily calcium supplements.

Bones are made through a process called ossification, or the calcification of tissue into hard, bony substance. As a fetus, a human is made of soft cartilage, which is slowly replaced by bone throughout childhood. (New parents are intimately familiar with the fontanelle, or "soft spot," on a baby's skull.) Bones provide us with the foundational structure of the entire body—giving us our shape and form and supporting our ability to move. Up until our twenties, new bone replaces old bone at a faster rate, aiding in increased bone density. Toward late middle to old age, production of new bone typically slows down and is unable to keep up with the loss of bone density. This can lead to bone diseases as well as loss of functionality, overall weaker strength, and other health problems.

Historically, measuring and evaluating bones and bone structures was used as a guise; instead of addressing social inequality, scientists relied on physical features or malformations of bones to make claims about an individual's or an entire race's social worth. The field of *anthropometry*, or the measurement of human bodies, has been used to understand physical variations between individuals. While anthropometry may be useful and productive for applications in ergonomics, the scientific innovation of industrial design, the use of measurements of human differences has a long racialized and racist history.

For example, differences in bone structure, including skull size and leg length, have been used to justify race-based discrimination and beliefs about racial superiority. Beginning in the early 1800s, mapping of the human jaw bone led to the conviction that Africans (then called negroes) were most closely related to animals and hence not civilized, as their jaws were perceived to be bigger and stronger than other races (Jahoda 2007). However, as the section below explores, development of bones carries with it the social stratification of individuals based on their access to nutrition, vitamins, health care, and exercise.

Brittle bones

Osteoporosis, a term meaning "porous bones," is a condition in which minerals in the bones—especially calcium—decrease, causing a loss of density and increased porosity. The more porous bones are, the more likely they are to break. Vulnerable areas include hips, wrists, and spines, which can also become misshapen, for example in the distinctive curvature or "stoop" of some elderly people. Often, people are unaware they even have brittle bones until they experience a fracture from a fall or a particularly violent episode of coughing that results in a snapped rib. The disease can be diagnosed by a bone density scan, although such technologies are largely unavailable in resource-poor regions of the world.

White and Asian women are more likely to suffer from osteoporosis than men and women in other racial/ethnic groups. As researched by biologist Anne Fausto-Sterling (2005, 2008), race and sexual differences in bone density demonstrate the interaction of biology, environment, and life-course factors. Women over 45 and those who are postmenopausal are especially at risk of fractures and breaks due to loss of bone density, whereas men under 45 are more likely to break their bones through sports and risk-taking activities.

Risk factors for osteoporosis beyond age and gender include smoking, small stature, family history, hormone problems such as thyroid disorders (which disproportionately also affect women), and dietary habits including mineral deficiencies. Osteoporosis, largely considered to be a "women's disease," thus amplifies gender differences in health and physiology, as well as in prevention and treatment. Women have long been told to take calcium supplements, but a 2010 study found that the supplements could increase the risk of heart attack. How are women to decide which "disease" to prevent first?

Hormone therapy (HT) is often recommended for older women (and increasingly men) in order to offset the physiological effects of aging. Yet such interventions are contested, with health advocates suggesting there are safer measures to improve women's health than pharmaceuticals. For example, exercise—especially resistance activities such as weight training and swimming—has been shown in numerous studies to increase muscle mass and bone density. A Canadian study found that strength, maximum oxygen uptake, and bone density were all positively correlated (Chilibeck, Sale and Webber 1995). Studies have also shown the benefits of exercise for older people; in addition to building strength, staying active can help to maintain balance (important in preventing falls), relieve pain, improve overall wellbeing, and decrease mortality.

Bones and global health

It is important to note that although osteoporosis is often an unfortunate side effect of aging, it is neither inevitable nor the only bone disease worth considering. It is, however, instructive. Just as poor nutrition and lack of exercise can lead to loss of bone density in older women, malnutrition the world over in children and young adults can also lead to brittle bones, as well as to organ failure, muscle weakness, poor immune function, and other serious health problems.

For example, rickets is caused by vitamin D, calcium, and/or phosphate deficiencies and can lead to weakening and softening of bones. The prevalence of rickets is overwhelmingly in the Global South, affecting children and creating skeletal deformations. Like rickets, osteoporosis allows us to see many ways in which health and illness are socially produced through lack of adequate resources such as food. Brittle and broken bones, like many other diseases, are stratified by gender, age, income, and geography.

Thus, analyzing bones can tell us much about global inequality.

Bone marrow transplants

Sickle cell anemia, as discussed earlier in the book, is a hereditary disease in which the body's red blood cells are crescent or sickle shaped rather than round. Because the red blood cells move oxygen through the body, the abnormal shape means that less oxygen is delivered to body tissues. The resulting anemia can lead to fatigue, bone pain, jaundice (yellowing of the skin), rapid heart rate, shortness of breath, and abdominal pain. More severe symptoms include bone infection, lung and urinary tract infections, loss of eyesight, ulcers, and arthritis. There is no known cure for sickle cell anemia, but treatment can prevent pain and other symptoms.

Sickle cell anemia, the most common genetic disease in the world, is inherited from both parents; if only one parent is affected, a child will carry the sickle cell trait. The disease is more prevalent in people of African and Mediterranean descent, although also affects those from the Caribbean, Central and South America, and the Middle East. The World Health Organization reports that approximately 5 percent of the world's population carries the genes for hemoglobin disorders, including sickle cell. The Sickle Cell Disease Association of America estimates that about 100,000 people in the United States live with sickle cell.

June 19 is recognized as World Sickle Cell Day.

Bone marrow transplantation has been used to treat sickle cell anemia since the mid-1980s. Autologous bone marrow transplant involves using the patient's own marrow; allogeneic bone marrow transplant uses marrow from another person. In either case, marrow must be harvested from the donor. Bone marrow transplantation is expensive, requires highly specialized care, and can be life threatening, making it much less viable in the Global South where health care resources are limited to nonexistent.

Where bone marrow transplantation is an option, up to 85 percent of patients who meet the criteria remain disease-free; however, only about 7 percent of sickle cell patients meet the criteria. These include being aged 16 years or younger, having severe symptoms but no neurological damage, and having a genetically matched sibling who will donate bone marrow. Approximately 10 percent of patients receiving bone marrow transplants die. Those who live face a higher risk for cancer and infertility.

Recent years have seen the advent of *stem cell transplantation* for sickle cell anemia, although as we discussed earlier, stem cell research is mired in politics and regulatory challenges in the United States and Europe. In some cases, stem cells are removed from umbilical cord blood in affected children and stored until they can be used in transplantation.

Bone marrow transplantation for sickle cell anemia raises a number of ethical questions, most significantly the costs versus benefits of treatment. Is it worth the risks to donors and recipients of bone marrow to attempt to "cure" sickle cell anemia? Who gets to decide if these risks are worth taking—children affected or their parents, who legally can make decisions about their health care? Also, because treatment is limited to countries with advanced health care resources and technologies, what are the global ethics of treating some children and not others? How do race, ethnicity, and geography factor into the promises of bone marrow transplantation technology? What does it mean that the highest prevalence of sickle cell anemia is in West and Central Africa, but that bone marrow transplant networks are located largely in the Global North?

Using our critical lens, we see that bones are not only the structural foundation of the human body but also a potential resource to be mined for treatment of disease. Living in a world stratified by race, class, and status, this resource is not universally accessible and brings into focus global inequality.

A CLOSER LOOK

Bones as evidence of trauma and genocide

The organization One Million Bones uses hands-on art making as a social practice to foster awareness about genocide. An installation in 2013 brought 1,000,000 handmade bones to the National Mall in Washington, DC. This symbolic mass gravesite, spearheaded by artist activist Naomi Natale, was inspired by atrocities of the Rwandan genocide and violent conflicts in Sudan and the Democratic Republic of the Congo.

In attempts to raise international awareness of genocide, this project hosts bone-making events, offers age-appropriate lesson plans for integration of bones into the classroom, and employs social media to stimulate interest and organize community art events. This form of activism is a type of embodied protest that uses corporeality of bodily suffering to redress social injustices. Here, bodies are simultaneously sites of resistance and reconciliation.

Genocide is defined as the deliberate killing, often through state-sponsored violence, of a large group of people presumed to be from the same ethnic, tribal, or cultural group. The Geneva Conventions—that is, statutes of the International Criminal Tribunal in The Hague—provide a global legal system for addressing war crimes. Yet proving war crimes, especially genocide, is difficult due to the ubiquity of global ethnic conflict and geopolitical implications of accusation. Although genocide may seem obvious with the discovery of mass graves, additional investigation is often required. In many instances, forensic anthropologists are called on to investigate bones at mass gravesites to determine if ballistic trauma is indicative of homicide by civilians or the result of military engagement (Warren 2007).

Just as bones are used to prove certain truths about war crimes, bones have also played a role in reconciliation, although not without controversy. While experienced as a catharsis for some visitors, the use of bones in genocide memorials in Rwanda and Cambodia provokes concerns about identity and consent of those whose bones are displayed (Guyer 2009; Sion 2011). There are a handful of places in the world, such as the Catacombs in Paris, where stacked-up bones serve both historical and tourist functions.

Displays of bones, no matter the intent, conjure the fear of spectacle, whereby individuals may visit bones not to grieve but to gawk, and may even take pieces of bone as souvenirs. What is the role of memory, grief, and collective mourning when bones are thus displayed for visual consumption?

Sporting muscles

Moving from bones to muscles, we continue our exploration of the cultural meanings of these foundational body parts. Whereas bones are often understood in the context of morbidity and mortality, as in osteoporosis or the forensic study of remains, muscles are more often understood in the context of growth, health, fitness, and strength—or lack thereof.

The classic 1977 documentary *Pumping Iron* features several male bodybuilders as they vie for the Mr. Universe and Mr. Olympia titles. The film offers an intimate look at grueling regimens to sculpt the body, as the men are judged on their muscular symmetry, clarity, size, and proportions. Focusing tight camera angles on the lats and delts of the primary character, soon-to-be celebrity Arnold Schwarzenegger, the film details psychological, physical, and interpersonal challenges bodybuilders face in their preparation for competition.

At one point, while at Gold's Gym demonstrating techniques, Schwarzenegger explains:

> The greatest feeling you can get in a gym, or the most satisfying feeling you can get in the gym is . . . The Pump. Let's say you train your biceps. Blood is rushing into your muscles and that's what we call The Pump. Your muscles get a really tight feeling, like your skin is going to explode any minute, and it's really tight – it's like somebody blowing air into it, into your muscle. It just blows up, and it feels really different. It feels fantastic.

Pumping, preening, and flexing his muscles proved to be an excellent training ground for Schwarzenegger's subsequent Hollywood fame and political success as hypermasculine action star of the *Terminator* films and Governor of California (2003–2011). His famous massive muscles influenced a generation.

As beautifully detailed in Sam Fussell's memoir about becoming a professional bodybuilder, the power of muscle to provide a sturdy and impenetrable masculinity is compelling. Fussell (1991: 25) shares an epiphany about his own fragility and the solutions to a more convincing masculine exterior he found while paging through images of Arnold in a bookstore:

> And that's where it hit me, right there in the Strand. I knew it in an instant, my prayers were answered. What if I made myself a walking billboard of invulnerability like Arnold? Why couldn't I use muscles as insurance, as a certain indemnity amidst the uncertainty of urban strife? Arnold had used iron to his obvious advantage, why couldn't I? And if the price was high, as a quick glance at the tortured faces in the training photos suggested, well, wouldn't four hours a day of private pain be worth a lifetime of public safety? . . . Big loud muscles – well, they were something else altogether. Surely a quick appraisal of my new gargantuan body would guarantee me immunity, even from the criminally insane.

As both Schwarzenegger and Fussell illustrate, muscles do things in social worlds: they empower and intimidate, provide status and security, and inspire envy or revulsion. And not just for men—*Pumping Iron* was followed by *Pumping Iron II: The Women*, which offered a look at female bodybuilders. These women challenged gendered stereotypes about how women should look, how big their muscles should be, and how strong they should be. Where some viewers saw beauty in the women's physiques, others were horrified that women could appear so huge and "abnormal."

For example, Bev Francis, an Australian power lifter and shot put champion, drew particular criticism for her "failure" to meet standards of femininity for women bodybuilders. As explored by psychologist Priscilla Choi (2003), there is a persistent and growing social and cultural resistance to excessive female muscularity as it is seen as a threat to the patriarchal social order. In Choi's research, female bodybuilders in competition "emphasize their sexuality [by] adopt[ing] sexual positions as part of their posing routine and as a result this emphasis incite[s] wolf whistling, lewd remarks, and other behaviors from the mostly male audience" (Choi 2003: 77).

The business of muscles

The idea of possessing toned, but not necessarily bulging, muscles has seeped into popular media in the Global North, spawning highly specified ideal body types for each gender and generation. Men need to be bigger and harder, while women need to be toned and flexible. Since the 1970s, shifting social norms about the sculpted body have created a correspondence between health, fitness, body size, and social worth. In some countries, the more fit you are, the healthier you are considered to be, and the more deserving of social respectability due to your diligence in being a productive citizen. The obverse is also true: if you are not fit (i.e., toned and flexible), you are not considered to be healthy nor are you recognized as a good citizen.

The growth of fitness magazines, the supplements industry, recreational centers, yoga and pilates practices, and fashion trends have driven many in the Global North to become acquainted, at increasingly specific levels, with our muscles. Many of us are familiar with the more common muscle groups such as triceps, biceps, glutes, and calves, but increasingly, we also can identify and talk about more obscure muscles or connective tissues, such as the psoas or iliotibial band.

This preoccupation with our muscles signals a shift from engaging in hard physical labor, as was the case historically for many in the Global North who now hold "desk jobs" and as is still the case for much of the Global South, who toil in agriculture, manufacturing, and other challenging sectors. Now, for many in the Global North, development of strength and mass is not simply a byproduct of hard work, but rather an object of desire, a self-conscious "body project" (Brumberg 1997).

The multi-billion dollar fitness enterprise deliberately overlaps with Hollywood's entertainment forces and mass media to shape cultural norms about desirable bodies, as well as products and methods to achieve these physiques (Tan and Brignole 2012). However, such connections between our "six-pack abs" and our social worth generally privilege those who *already rank* more highly in systems of social stratification, given the amount of time and capital necessary to sculpt the human body deliberately.

As sociologists Shari Dworkin and Faye Wachs (2009: 23) argue: "the ways in which signifiers of 'health' and 'fitness' come together to mark moralities, privilege certain lifestyles, and exclude others in a given contemporary movement are even more meaningful given that the messages attached to images and ideals are often conflated with a state of health in the name of science."

We attribute symbolic and actual value to muscles in that they represent a certain type of lifestyle, ethic of self-care, and contribution to the social body. In fact, we have developed entire professional fields and academic disciplines to study the movement of bodies and muscles including dance studies, chiropractic medicine, and physical and occupational therapy. Exploring one field in detail, we turn our attention to kinesiology.

Kinesiology

There are about 650 to 800 muscles in the human body, depending on how distinct muscles are categorized as either individual muscles or in groups. Altogether, muscles make up about 40 percent of our body weight.

Divided into three groups, cardiac muscles are found only in the heart, where they continuously pump blood throughout the body. Smooth muscles are controlled automatically by the central nervous system (CNS); since we do not consciously think about flexing our arterial walls or intestines, these muscles are termed involuntary. Skeletal or striated muscles, perhaps the most familiar, are attached to bones, and we can control these muscles to do things, such as lift a box, or move our bodies to and fro.

For centuries, the magnificence of humans' physical strength and flexibility has been reflected in poetry, sculpture, athletics, literature, music, dance, and games. Beyond artistic endeavors, admiring and understanding muscles is the bailiwick of an interdisciplinary field called *kinesiology*, or the study of the movement of human bodies.

An applied kinesiologist is an expert in muscle tension; however, kinesiology has come under attack as a form of pseudoscience by biomedical allopathic providers, whereby muscle testing has been labeled quackery by allopaths (Atwood 2004). Massage therapy is a form of healing that targets muscles and is widely used in the Global North by people with economic resources and/or some types of health insurance. Massage, along with chiropractic medicine and other "alternative" practices, is becoming more recognized as critical to overall wellbeing. Massage and touch are important features of some religious and spiritual practices, particularly in Asia.

A kinesiologist is skilled at determining how bones, muscles, ligaments, and tendons work in tandem to create an integrated body. Muscle testing can reveal structural or chemical imbalances in the body. Often, these tests are not considered legitimate in biomedical settings, because surgeons and sports medicine physicians are typically considered experts at muscle and bone health. However, kinesiology has garnered intellectual and professional acceptance in other fields. For example, in the social sciences, kinesiology has contributed to the growing field of physical cultural studies, a contextual understanding of corporeal practices, discourses, and subjectivities through which active bodies become organized, represented, and experienced in relation to the operations of social power.

Gender, sport, nation

For critical scholars of the body, muscles are significant because they straddle both the individual body, as each person attempts to perfect their musculature in light of specific cultural norms, and the political body, as states create programs to present a strong national image through athletic competitions, military preparedness, and economic policy. Muscular bodies may convey local, state, and national pride, as in the case of World

Cup or Olympic athletes who represent symbolically the fittest and the best, or in the adoration showered on football, soccer, and baseball teams.

The sociology of sport examines intersections of sport and society, and in particular how sport is made meaningful in people's individual, institutional, and political lives. In addition to cultural analysis of muscularity and performance, sociologists of sport also examine concerns about oxygen uptake, embodied processes, enhancements, and the role of human metabolism. Through the perspective of this subfield, we can see how muscles matter in transnational, global, and other macro-sociological contexts (Silk, Andrews, and Cole 2005).

Sports, understood as more than just leisure activities for creating personal fitness, are also a multinational and highly lucrative business and spectator enterprise. A mechanism of upward mobility transcending class stratification for the very few, the ideology of athletic achievement is that it comes with glory for the winner and their homeland, but also offers a big payday and mega stardom. Social consequences ensue when lucrative possibilities for athletic performance are broadcast and internalized.

Similar to a brain drain, where highly educated and skilled labor leaves nations of economic instability for more lucrative sites, economist Wladimir Andreff (2006) has coined the term *muscle drain* to refer to international transfer of athletes from a "less developed" country, particularly on the African continent, to "more developed" countries. This loss of talent deprives the country of origin of the opportunity to populate sports programs that would enable them to compete with other nations, while simultaneously enabling the highest bidder the chance to curate a talented field of international athletes who have little connection to the country.

For example, currently, Qatar fields Brazilian-born soccer players in the FIFA World Cup. Cuban-born triple jumper Yamile Aldama competed for Sudan in the 2004 Olympics and the British team in the 2012 games.

Gendered muscles and bones

As indicated in the beginning of this section, muscles are especially important for the expression and affirmation of a solid and strong masculinity. Therefore, when women have muscles that are outside of the culturally expected range (i.e., too large), their gender is called into question. So, too, with men who do not have large muscles; they are often considered effeminate, and may be harassed and even assaulted by larger men for failing to be "manly" enough.

People are thus divided on the basis of physiology—musculature, bone density, endurance, and other physical abilities. But are human males and females so clearly differentiated on the basis of physical ability? We behave as if they are, because when other criteria for competition are introduced—people in wheelchairs competing against each other; amputees competing against each other and even against able-bodied athletes; weight classes, age groups—these groups, too, are divided into two different gender categories.

Girls and boys may play on the same teams when they are young, but not when they reach puberty. Indeed, in many contexts there are greater school-based sporting opportunities for boys than there are for girls. In professional sports, in major collegiate competitions, and in the Olympics, the rule is that women do not compete against men. The rationale, of course, is that it would not be fair because women would always lose. But would they? We may never know, because they rarely compete publicly against men.

Pursuing growth in muscle mass and athletic achievement, even among highly trained athletes, is a challenge of balancing traditional social expectations of masculinity and femininity with intense physical training and changing body structure. Sociologists Mari Kristin Sisjord and Elsa Kristiansen (2009) examined the conflicts experienced by Norwegian elite women wrestlers in their pursuit of athletic achievement. The athletes' performance of femininity, expected socially of women, is compromised by transformations in their physiques and in athletic practices—just as Bev Francis's was in *Pumping Iron II*. Female bodies are expected to look and behave a certain way, and when they do not, there may be a high cost.

The International Olympic Committee dropped gender verification tests for women in 2000. In 2004, it ruled that male-to-female and female-to-male transgendered people could compete in their new gender, provided they had evidence of "appropriate surgery," had completed hormone treatment at least two years prior, and were legally recognized as members of their new gender.

A highly contested issue in recent years concerns athletes with *intersex conditions*, such as "masculinizing disorders" that are perceived to offer female athletes with higher levels of testosterone an unfair advantage. In 2010, the IOC suggested that athletes with intersex conditions undergo hormone treatment—a contested recommendation given the long history of pathologization and medicalization of people with intersex conditions (Karkazis 2008; Reis 2012).

FIGURE 6.1 Zulfiya Chinshanlo
Photo courtesy of Rob Macklem Zulfiya Chinshanlo of Kazakhstan, a world record holder in weightlifting, competes the 2009 53 kg class at the Goyang City site of the World Championships. She also competed in the women's 53 kg weightlifting event at the Olympic Games in London on July 29, 2012. She was born in China. What is your first reaction to this image? What do you think of the gender of this lifter? How is the display of her muscles "read" in your culture?

A CLOSER LOOK

Breaking records, breaking rules

In 2009, when Caster Semenya, an 18-year-old South African woman, won the 800-meter race at the World Championships in Athletics in Berlin, her womanhood was challenged by one of her competitors, and a *gender verification test* was called for (Clarey and Kolata 2009). Semenya won with a time of one minute, 55.45 seconds, the best in the world in 2009, beating the defending champion by 2.45 seconds. Her time was not the all-time fastest, yet none of the other champions had to prove they were women. Semenya was described in one news report as having "an unusually developed muscular frame and a deep voice" (Kessel 2009), hardly accurate criteria for gender verification. The sprinter was subsequently reinstated and cleared to compete against women.

However, trying to develop "simple rules for complex gender realities" presents a major challenge to gender segregation, wrote Alice Dreger, professor of clinical medical humanities and bioethics at Northwestern University (2009). There is no direct correlation between a genotype and the complex manifestation of a social "phenotype" such as gender. Measuring testosterone levels might give an indication of muscle mass, but then should men with low testosterone levels compete as women and could women take testosterone boosters to compete with men, who are allowed to boost their levels if they are "low"?

Caster Semenya subsequently appeared on the cover of *You* magazine, looking decidedly more "feminine." This suggests that no matter how talented the athlete, gender rules still apply—not just in terms of how women perform, but how they look. The image of Semenya is deployed in ways to illustrate her claim to her gender identity. Is this equality? Here are some images of Caster Semenya as a competitor and as a cover girl:

FIGURE 6.2 Caster Semenya
© AP Photo/Lee Jin-man, AP/Press Association Images

FIGURE 6.3 YOU Magazine cover
Photo courtesy of YOU Magazine

Promises and perils of enhancement

The pressure to build muscle and perform superhuman athletic feats fuels an industry of pharmaceutical enhancements that promises to increase muscle production and improve physical performance. International anti-doping agencies in amateur and professional sports use the term "performance enhancing drugs" to refer to anything that improves upon the body; this can include stimulants, sedatives, and blood boosters.

Anabolic-androgenic steroids (AAS), available as injectables, creams, gels, or pills, are manufactured substances related to male sex hormones (e.g., testosterone). Anabolic refers to muscle-building and androgenic refers to increased male sexual characteristics. The term "steroids" refers to the class of drugs.

There are several so-called legitimate uses for AAS that require a prescription: for treatment of breast cancer, testicular cancer, or for pituitary malfunction, to name a few. However, there is also a growing market in *steroid abuse* despite their ban by all major sporting organizations. Through international smuggling and clandestine labs, drugs are created and distributed in underground marketplaces. As a result, it is difficult to find an accurate figure on the prevalence of steroid abuse in professional sports.

One thing is clear, though, and that is how steroids traffic in young adults' lives. In the United States, the National Institute on Drug Abuse (NIDA) claims that more than half a million eighth- and tenth-grade students are currently using steroids without knowledge of the health risks. Steroid use by young people is greater among males and those participating in at least one sport (Lorang et al. 2011). In Australia, which has some of the strictest laws in the world against steroids, seizure of steroids and arrests are the highest on record, and the number of steroids and trafficked pharmaceuticals detected at the border is also at very high levels.

Steroid abuse can lead to serious, even irreversible health problems: liver damage, jaundice (yellowish pigmentation of skin, tissues, and body fluids), fluid retention, high blood pressure, increases in LDL (bad cholesterol), and decreases in HDL (good cholesterol). Steroids are also linked to behavior changes including increased irritability and aggression. Ironically, among the possible consequences for men of steroid abuse are feminizing physiological changes such as the shrinking of the testicles, reduced sperm count, infertility, hair loss, and development of breasts.

Adolescents, whose bodies are still developing, risk stunted growth due to premature skeletal maturation and accelerated puberty changes; adolescents risk not reaching their expected height if they take AAS before the typical adolescent growth spurt. Here, we see a variety of health problems caused by the pursuit of "health" in the form of musculature. Gender dynamics are at play, as are prevailing standards of attractiveness—cultural messages to which young people may be especially susceptible.

Perhaps the most infamous case of steroid doping in recent history is that of Lance Armstrong. The famous U.S. cyclist is as widely known for winning seven Tour de France titles as he is for beating testicular cancer. Now, he will likely be equally recognized as the man who lost seven titles due to illegal use of performance enhancing drugs.

Armstrong had long dodged allegations and denied rumors that he used drugs; indeed, he used his cycling success, cancer survival, and nonprofit foundation to maximize his "American hero" image (Casper and Moore 2009). In 2012, the U.S. Anti-Doping Agency found him guilty of using drugs, and also of instituting elaborate strategies repeatedly to pass drug tests. This finding was supported by the Union Cycliste Internationale.

Consequently, Armstrong was banned from professional cycling for life and stripped of his Tour de France titles, and he also lost a number of lucrative endorsement deals.

Oscar Pistorius and "technological doping"

Oscar Pistorius, nicknamed "Blade Runner," is a South African athlete who is also a double amputee. He was born in 1986 without fibula—the long bone that runs from the knee to the ankle—in both of his legs. His parents made the decision to amputate his limbs below the knee. He received his first pair of prosthetic legs six months after the surgery.

Raised in a "sports-mad" family, Pistorius engaged in many sports including water polo, rugby, tennis, cricket, wrestling, and boxing. In 2003, after shattering his knee playing rugby, he began to run as part of his rehabilitation. And he has never stopped running, winning numerous races and breaking world records. Sporting the Flex-Foot Cheetah, he is a distinctive figure on the track.

Considered "disabled" because of his amputations and prosthetic use, Pistorius drew controversy when he made a bid to run in the 2012 London Olympics alongside able-bodied athletes. Recognized as one of the fastest sprinters in the world, he competed in the 400-meter and the 4x400 relay in the 2012 Games. He did not win, but he made history.

The extensive investigation by the Olympic officials at the 2012 Games concluded that the runner's prosthetics did not give him an unfair advantage in competition. Why were so much research, testing, and testimony required to decide his participation? If carbon fiber prosthetic legs do not give an unfair advantage, serving as a kind of "*technological doping*" to improve his odds, are there interventions to bones and muscles that could? Or are such interventions simply an extension of natural, hard-earned athletic prowess on the track? Can human beings be both "disabled" and "able-bodied"?

Subsequent to the 2012 London games, Pistorius drew another kind of controversy. In 2013, he was formally indicted for the premeditated murder of his girlfriend, Reeva Steenkamp (Imray 2013).

Conclusion

Understood anatomically, muscles and bones are brought together by tendons, also called sinews. This strong fibrous cord is extremely tough yet flexible, and contracts and expands depending on the direction and force of muscles and bones. In order to perform any voluntary movement, we rely on exquisitely choreographed interrelationships among tendons, bones, and muscles. We need bones and muscles, working in tandem, to jump, walk, dance, lift, run, hug, somersault, and engage in a variety of other movements. When they do not work, we may be disabled from participating fully in sport and other activities.

As critical scholars of the body, our job is to examine linkages or *social tendons* connecting the muscles and bones of the physical body to processes of socio-cultural signification. Through these social tendons connecting the material body to the larger social world, we ascribe various meanings to muscles and bones. When we move through the world, similar to the coordination of muscle, bones, and tendons, our bodies interact with others in domains such as medicine and sport, revealing and even amplifying our differences and inequalities.

There are entrenched beliefs that the body—its bones and muscles—holds the key to understanding racial and gendered differences in humans. However, as this chapter has explored, human individuals and institutions labor diligently to ensure that bones and muscles develop in precise ways that maintain the gender binary, suggesting that differences are not as "natural" as historically presumed. Bones and muscles are also mined for information or performances that conform to already existing beliefs about race and gender.

National identity is often riding on the embodied performance of a muscle-bound man or not-too muscle-bound woman who can bring home the international glory. Individual athletes, such as Lance Armstrong and Oscar Pistorius of the United States and South Africa respectively, brought both glory and shame to their countries.

See also

Cells and tissues; Organs; Circulation; Respiration; Endocrine system and metabolism

Key terms

Structure, anthropometry, osteoporosis, sickle cell anemia, genocide, kinesiology, muscle drain, intersex conditions, gender verification test, anabolic androgenic steroids, steroid abuse, technological doping, social tendons

Discussion questions

1 Should we prevent/treat osteoporosis with hormone therapies or emphasize other measures, such as exercise and good nutrition? If the latter, what are some ways more fully to integrate healthy eating and physical activities into daily routines of the elderly? How might nutrition be improved for all people globally to prevent brittle bones and other health problems?

2 Bones are central to many death rituals. Humans often bury their dead, which eventually, with the loss of flesh and muscle, become bone, while other species such as elephants have been seen "grieving" at the bones of their dead kin. Discuss the relationship between how bones are displayed or used and the "appropriate" social responses to these uses. Does the method of bone disposal or display matter to processes of mourning?

3 What do you think is the most socially desirable presentation of muscles for a man? For a woman? How does this desirable expectation change with respect to a person's age? By what criteria do we define a "good" muscle mass?

4 Cyclists Lance Armstrong and Floyd Landis, sprinter Marion Jones, sprinter Ben Johnson, and professional baseball player Barry Bonds each have had their careers tarnished by illegitimate use of steroids. What is your opinion regarding the use of steroids? What is at stake for athletes who test positive? In your view, what are some social, economic, and personal conditions that may lead to use of steroids, despite their social stigma and illegality?

5 Can you name some contemporary examples of muscle drain in professional and semi-professional sports? Given this muscle drain, what does it mean for a nation to say they have the best athletic team? If muscle drain persists, what do you forecast about global athletic competition?

PART II

Systems and their representation

chapter 7

Central nervous system

Introduction

In 1990, U.S. President George Bush proclaimed the Decade of the Brain, inaugurating a host of programs, activities, and funding opportunities. This bold announcement linked brain research to the U.S. War on Drugs, including the effects of drug and alcohol use by pregnant women on their unborn fetuses. The Proclamation thus embedded the call for a "new era" of scientific research within a conservative political agenda emphasizing fetal rights (Roth 2003). By the mid-1990s, there was some concern that not enough was being done to realize the ambitious goals set out in the Proclamation (Goldstein 1994).

Still, research proceeded apace. Recently, the European Brain Council launched a project to make 2014 the Year of the Brain in Europe, and U.S. President Barack Obama announced a multi-million (and potentially multi-*billion*) dollar investment in a coordinated project to map the human brain (Markoff 2013). The immensity of this undertaking ranks it among the most ambitious scientific efforts in history, and its implementation promises to influence how we consider the central nervous system. The project is tied to the search for cures for diseases such as Alzheimer's and Parkinson's, as well as to advances in artificial intelligence.

The 21st century has ushered in what many are calling the Age of Neuroscience (Feilden 2012). From neuroethics to neurogenics, neurosociology to neurotrauma, attention has shifted to the central nervous system (CNS)—with the brain still firmly in charge—as a blueprint for human behavior, capable of explaining everything from addiction to depression, love to war (Cozolino 2006). Brains, it seems, are both inside of us and all around us, and "neuro" has come to stand in for all things brain-related. Increasingly, we look to the brain as the source of human social life, although this privileging of the "neuro" is not without consequences or critics (Satel and Lilienfeld 2013).

This chapter explores the CNS in its social, cultural, and political contexts, including the brain's starring role in Big Science. We first describe the central nervous system and its component parts, then move on to a consideration of what "neuro" means in the 21st century. We offer "A closer look" at the search for gender and racial differences in brains, and then discuss CNS disorders including research initiatives, treatment, and global health inequities. We next focus on spinal cord injuries and traumatic brain injury. Last, we consider mental health, neurodiversity, and pharmaceuticals, followed by "A closer look" at the Mad Pride movement.

What is the CNS?

The central nervous system (CNS) is comprised of the brain and spinal cord. It is connected to the peripheral nervous system (PNS) via ganglia; nerves from the PNS attach to and from the central nervous system. What is "central" about the CNS is that it is the hub of integration and coordination of the entire organism in most species (Brodal 2010). It is, to borrow a military metaphor, "command central" for the body's functions and movement.

The central nervous system is typically well protected, although it may be susceptible to damage of various kinds. The brain is contained within the cranial cavity and, in vertebrates, surrounded by the skull. The spinal cord is contained within the spinal cavity and protected by vertebrae. Both are suspended in fluid. The spinal cord connects to the base of the skull, and nerves project from the spinal cord to other parts of the body (Nieuwenhuys, Voogd, and van Huijzen 2007).

An adult human brain weighs about 1.5 kilograms (about three pounds), although there may be significant variation among individuals. The largest part of the brain is the cerebrum, made up of two cerebral hemispheres. These hemispheres are covered by the cerebral cortex, which is divided into four lobes (frontal, parietal, occipital, and temporal). The brain stem is underneath the cerebrum, and at the back of the brain is the cerebellum (Carter 2009). Like the brains of other mammals, the human brain is comprised of three parts: the forebrain, midbrain, and hindbrain.

The brain is a complex organ, made up of cells, neurons, blood vessels, tissue, and other matter. Tomography, a technique for imaging sections of matter, has shown that the human brain contains about 200 billion

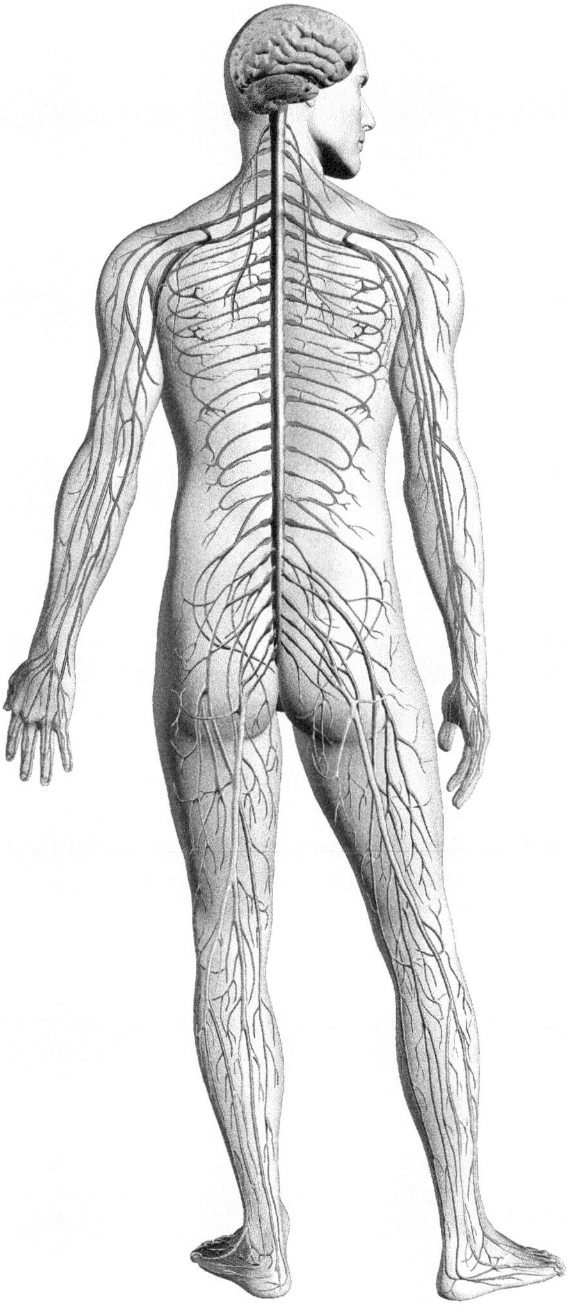

FIGURE 7.1 Human nervous system
© Shutterstock

neurons, and that just in the cerebral cortex, there are 125 trillion synapses (Micheva and others 2010). Different parts of the brain are responsible for different functions in the body. Some parts of the brain may become damaged, while others remain intact. As we discussed in the "Organs" chapter, "brain death" can occur while other body parts, including viable organs, continue to function.

Neurons are specialized cells found throughout the nervous system that transmit signals as electrical current. *Synapses* are connections between neurons where the electrical signal must be transmitted through a neurotransmitter (e.g., serotonin and dopamine) that is released to the space between the two cells.

While historically brains were understood to be fairly static once children reached a certain age, now brains are understood to be dynamic and evolving across the life course. The term *neuroplasticity* is used to describe changes in brain structure and function due to environmental factors, physical and intellectual activities, injury, and disease (Doidge 2007). For example, the brain's neuroplasticity underlies theories about trauma (Kolassa and Elbert 2007), memory (Bermúdez-Rattoni 2007), addiction (Hyman and Malenka 2001), and training for brain health (Draganski et al. 2004). The brain's plasticity has been taken up in popular culture with exhortations for everyone, but especially the aged, to "keep their mind sharp" through puzzles, mental exercise, and physical fitness.

The spinal cord is also complex. Although many people think of the entire vertebral column as the spinal cord, the cord extends only from the occipital bone to the space between the first and second lumbar vertebrae. Its main function is to transmit neural signals between the brain and other parts of the body, but it also coordinates reflexes. Spinal cord injuries can lead to a disabling loss of vital functions, including movement (Wyndaele and Wyndaele 2006).

The central nervous system, with its primary functions of command and control, is frequently used as a descriptor for social life. We often refer to the "brains" of an operation, or even "she's the brain of the operation." Neural networks, a key component of *artificial intelligence*, have long captivated the public. Artificial intelligence is the use of a computer or automated system to perform actions thought to require intelligence, such as logical problem solving, creativity, empirical decision-making, and linguistic comprehension. For example, sociologist Jon Guice (1998) studied the history of defense funding and neural networks, a connection seen in the popular film *The Terminator* and its sequels, which relied heavily on the science of neural nets. From science to science fiction, humans have evidenced a deep fascination with the neural basis of our existence.

Living in the "neuro" age

Neuroscience is the field of study of the nervous system. With roots in Ancient Egypt, the field has evolved over the centuries from trepanation (drilling a hole in the skull), through phrenology (Greenblatt 1995), to modern neuroscience founded on molecular biology and other advances. As with many sciences, development of the microscope, including the ability to take apart and visually examine pieces of the brain, helped to develop the field. Neuroscience, a broad field encompassing everything from basic research to clinical application, is currently enjoying considerable prestige and authority (Weisberg et al. 2009).

Consistent with the rise of the "neuro," we have seen the emergence of *neuroethics*, a field devoted to examining the ethical, legal, and social aspects of neuroscience (Roskies 2002). We've also witnessed the emergence of neurogenetics, which looks at the role of genes, genetics, and the human genome in the nervous system (Simón-Sánchez and Singleton 2008). Neurolinguistics explores the ways in which language is developed through neural processes; neuroaesthetics looks at the neural basis of how we look at and appreciate art and other visual media; neuropsychology explores the underlying neural basis of psychological behavior; and neurosociology examines how biological processes and neural networks shape social life (Franks 2007).

In all of these areas, the prefix "neuro" references a biological story we tell about ourselves as humans. The use of the "neuro" is similar to the use of genes and genetics, as discussed earlier in the book, to explain all human behaviors and evolution. In the "neuro" version of the human story—which increasingly is becoming the hegemonic story—we are our brains. Philosopher Catherine Malabou (2008: 1) writes: "The brain is a work, and we do not know it. We are its subjects—authors and producers at once—and we do not know it." Sociologist Nikolas Rose and historian of science Joelle M. Abi-Rached (2013: 1) write: "It seems that these other ways of thinking of ourselves—of our psychological lives, our habitual activities, our social relations, our ethical values and commitments, our perceptions of others—are being reshaped. They must now be grounded in one organ of our bodies—that spongy mass of the human brain[.]"

In short, living in the "neuro" age means that various dimensions of our entire existence are reduced to our brains, making certain stories possible while also foreclosing others (Quart 2012).

A CLOSER LOOK

Gender, race, and the brain

As we've discussed in previous chapters, the history of science narrates an ongoing search for gender differences between men and women, and among different ethnic groups and races, on the basis of anatomy. Scientists from different fields have attempted to locate sexual and racial differences in various body parts, such as bones, (Fausto-Sterling 2005, 2008), hormones (Oudshoorn 1994), and the human genome (Richardson 2013).

However, feminist scholars have demonstrated that such efforts are *a priori* shaped by cultural understandings of gender and racial differences (Haraway 1989). That is, scientists already understand there to be differences and thus seek to "discover" the truth in ostensibly objective scientific research. As historian of science Sarah Richardson (2013: 13) argues: "gender is formative to the knowledge, practices, and institutional structure of the sciences."

The late evolutionary biologist and historian of science Stephen Jay Gould (1981), for example, described incidents of *scientific racism*. He discussed the efforts of figures such as Paul Broca, a 19th-century French physician and anatomist, and Samuel George Morton, a 19th century American physician and naturalist, to measure skull size among "Caucasians" and "Africans," and thus to identify differences in intelligence. Gould suggested that not only were the scientists biased from the start, but they falsified their results in favor of smaller skull sizes and lesser intelligence among Africans.

Psychologist Carol Tavris (1993: 44) took on the quest for sexual differences, suggesting that "when scientists haven't found the differences they were seeking, they haven't abandoned the goal or their belief that such differences exist; they just moved to another part of the anatomy or a different corner of the brain." She cites French social psychologist and anthropologist Gustave Le Bron, writing in 1879, who wrote of women's inferiority as "so obvious that no one can contest it for a minute." As with Broca and Morton, when the scientists in Tavris's study failed to find differences, they simply recalibrated their methods to make the data speak what they wanted to hear.

In the 20th century, efforts to inscribe gender differences in the brain centered on "right brain/left brain" distinctions derived from lateralization theory (Tomasi and Volkow 2012). Following from Paul Broca and others, scientists understand that some cognitive functions are controlled by one side of the brain, while others are controlled by the other side. People are presumed to be more "emotional" or "rational," and thus comparatively better at certain tasks than others, depending on whether they are right-brained or left-brained. Although the theory has been challenged (Nielsen et al. 2013), there remains a popular conception that such differences are real and influence gender roles (Jarrett 2012).

Given the persistence of gender and racial inequalities globally, the scientific search for differences in the body and its parts is not likely to abate anytime soon. However, sociologically, we can and should ask what meanings and consequences these differences, once "discovered," hold for the social arrangement of bodies. As long as differences translate into inequalities, the "facts" of anatomy warrant interrogation.

Diseases of the CNS

There are a number of diseases and disorders that affect the central nervous system, including functional disorders, degenerative conditions, and infectious diseases. From encephalitis to epilepsy, migraine to Huntington's disease, many of these carry a heavy burden physically and economically. We discuss some of these in depth as a way to highlight issues related to structural inequality and global health. It is worth noting that while some diseases have attracted considerable funding and support, others have not; these differences do not necessarily reflect the actual distribution of disease, but rather who is able to speak for those affected in the nonprofit industrial complex (Samimi 2010), a system by which organizations must professionalize and compete for government and foundation funding.

Alzheimer's disease, which can cause dementia, affects nearly 36 million people worldwide, according to the World Health Organization (2012). It is neurodegenerative; that is, the neural system breaks down, causing malfunctions throughout the body. With Alzheimer's disease, cognition is affected such that people who suffer from the disorder lose their intellectual capacities and memory. Positively associated with aging—that is, older people tend to get Alzheimer's—everyday activities may be severely impacted, and people with dementia require specific types of care, with particular challenges in this regard in the Global South where resources are limited (Dementia Research Group 2004). Comparatively, in 2012, the U.S.-based Alzheimer's Association received US$249 million in revenue from donors.

Parkinson's Disease is a movement disorder understood to be caused by insufficient amounts of the neurotransmitter dopamine in the brain (Bernheimer and others 1973). People with Parkinson's experience trembling;

stiffness in their arms, legs, and trunk; slowness of movement; and problems with balance and coordination. In more severe cases, they may have difficulty walking, talking, and completing basic tasks. According to the European Parkinson's Disease Association, approximately 6.4 million people have Parkinson's worldwide. Like Alzheimer's, Parkinson's tends to affect people aged 60 and over, but young people can also be affected. The American television and film star Michael J. Fox was diagnosed with young-onset Parkinson's in 1991, and subsequently created the Michael J. Fox Foundation for Parkinson's Research, with total assets in 2012 of approximately US$106 million.

Amyotrophic lateral sclerosis, also called ALS or Lou Gehrig's Disease (named after the American baseball legend who died at age 37), is a disease that attacks neurons in the brain and spinal cord, specifically those affecting voluntary muscles (Wick 2012). Initial symptoms might include trouble walking, running, and writing, or speech problems. As the disease progresses, those affected lose strength, are unable to move, and are unable to breathe on their own. Most people with ALS die of respiratory failure. More men than women are affected, and certain geographic areas are particularly affected, including the Mariana Islands, New Guinea, and southern Japan. Globally, ALS is relatively rare but carries a heavy economic and social burden (Chiò and others 2013).

Meningitis is an inflammation of the tissue surrounding the brain and spinal cord. Viral meningitis is the most common form. Bacterial meningitis, due to both pneumococcal and meningococcal infections, is less common in the Global North but more dangerous; it can lead to stroke, brain damage, and death. In some parts of the world, meningitis is epidemic; sub-Saharan Africa is known as "the meningitis belt" (Peltola 2001). Even with antibiotics, up to 10 percent of people with meningitis there die. Survivors may suffer hearing loss, brain damage, and learning disabilities.

The meningococcal vaccine, which is required in many states in the U.S. before children can enroll in public schools, can prevent four types of meningitis. The vaccine is recommended for children, teenagers, and some adults including military recruits, people without a spleen, microbiologists exposed to the causal pathogen, and people traveling to countries where the disease is common. However, the vaccine is not available everywhere (Cohn and Harrison 2013). Securing adequate vaccination is an effort of several Global North NGOs focused on health issues in the Global South, such as PATH and the Bill & Melinda Gates Foundation.

Spinal cord injuries

Spinal cord injuries typically result from trauma to the spine in which a blow fractures the vertebrae, such as accidents, sports, falls, and violence.

Some injuries can actually cut through the spinal cord, but most typically, pieces of bone penetrate and tear the spinal cord or the vertebrae become damaged and press into the nerves. With such injuries, signals are not carried between the brain and the spinal cord, thus affecting the body's function. A complete spinal cord injury can result in paralysis. A person suffering an incomplete injury may be able to feel movement below the injury.

Spinal cord injuries can be disabling. People who have suffered such injuries may be paraplegic, which means paralysis or lack of motor or sensory function in the lower half of the body. Quadriplegia, also called tetraplegia, describes partial or total paralysis from the neck down. Quadriplegics experience both sensory and motor dysfunction. With paralysis, limbs may be affected, but so too are bodily functions such as waste, digestion, sexual function, and even breathing. Most people who experience paraplegia or quadriplegia are reliant on various technologies, including mobility aids such as wheelchairs and assistive living devices.

Spinal cord injuries occur worldwide, although there are countrywide differences in etiology (origin of the injury), male-female differences, age distributions, and complications (Ackery, Tator, and Krassioukov 2004). Such injuries are expensive to treat, and it is clear that people with fewer economic resources experience greater difficulties following a spinal cord injury. Some 80 percent of the world's one billion disabled people live in the Global South. Where resources are limited, mobility and assistive living devices may not be as widely available as they are in more affluent parts of the world (Borg, Lindström, and Larsson 2009). Chronic care may be unavailable, and people with disabilities often live "precarious lives" (Chouinard 2013).

Spinal cord injuries exemplify many issues faced by people with physical disabilities, from access to care and assistive technologies to social and economic marginalization. The world over, disabled people are more likely than nondisabled people to live in poverty and to experience violence and discrimination (Braithwaite and Mont 2009). For disabled people who are also part of other marginalized groups—such as women, people of color, immigrants, children, and the elderly (Rosenberg and Wilson 2000)—the challenges may be even greater.

The academic field of disability studies, which emerged from the disability rights movement (Charlton 1998), challenges normative assumptions about ability. Disability is framed not as an inherent problem of the body, but rather as a problem of social structure, environment, resources, and access (Davis 2013). In many environments, the world is molded to fit normative beliefs about how human bodies should behave and function with little regard for the range of variations in bodily morphology.

Traumatic brain injury

Traumatic brain injury (TBI) is an injury to the brain that disrupts normal brain function. From "mild" TBI such as concussion to "severe" TBI resulting in unconsciousness and permanent neurological damage, traumatic brain injuries are caused by blows to the head. They can occur as a result of sports contact, military conflict, accidents, falls, and violence, such as domestic abuse. Effects of TBI can be short-lived or chronic, and diagnostic categories are fluid. Historian Jennifer Terry (2009) describes TBI as a "signature wound" for our time.

Globally, approximately 10 million people are affected by TBI, and the World Health Organization estimates that it could become a leading cause of death, surpassing other diseases, by 2020 (Hyder and others 2007). The International Initiative for Traumatic Brain Injury Research (InTBIR) is a collaboration among the European Commission, the Canadian Institutes of Health Research, and the U.S. National Institutes of Health. Established in 2011, its purpose is to improve outcomes and lessen the global burden of TBI by 2020. WHO groups traumatic brain injury as a type of neurotrauma and notes that such injuries can lead to death and permanent disability. Obviously, where treatment options are limited, death rates are likely to be higher.

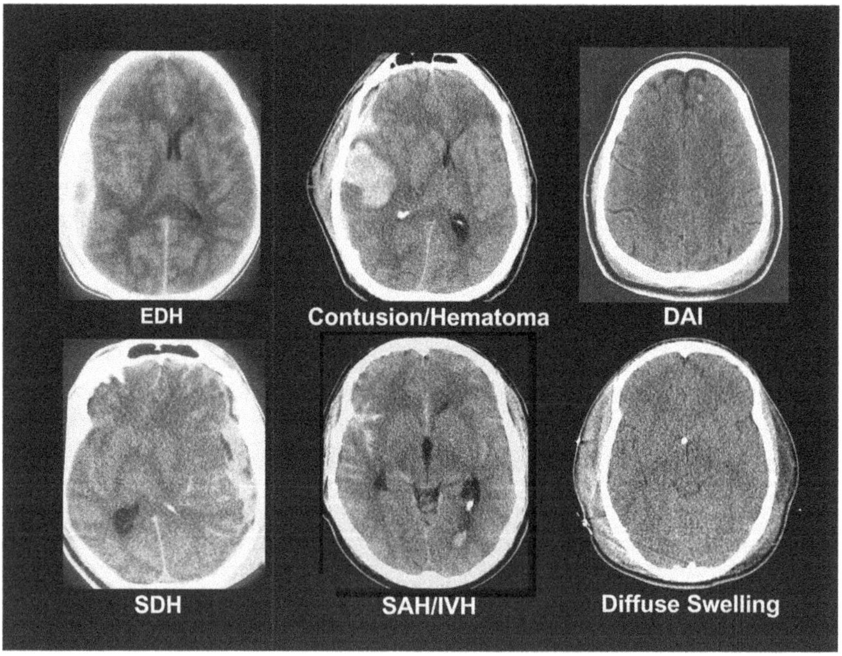

FIGURE 7.2 Brain scans
© Mary Ann Leibert Publications/American Psychiatric Publishing Inc.

Traumatic brain injury has received a great deal of attention in recent years. A number of sporting activities, both amateur and professional, pose the risk of concussion and severe TBI. Heading soccer balls, being tackled by a large defensive end in football, being hit by a hockey or lacrosse stick, crashing a motorbike or race car, skiing and snowboarding, diving, equestrianism, gymnastics—any and all of these pose risks of brain and other injuries. However, not all of these sports are equally lucrative, nor are they equally popular. In the United States, for example, considerable controversy surrounds the National Football League, a multibillion-dollar industry, and its resistance to mounting evidence of TBI among players and former players (Fainaru-Wada and Fainaru 2013).

Working with sociologist Daniel R. Morrison (2012), Monica has written about militarized traumatic brain injury as an exemplar of connections between disability studies and critical trauma studies. Media coverage of TBI among U.S. veterans has focused on two themes: IEDs (improvised explosive devices) and their role in causing closed-brain injuries, and "heroic" stories of veterans overcoming brain injuries. There is slippage in cultural and biomedical accounts between TBI and post-traumatic stress disorder (PTSD). Both diagnoses are used to explain violent episodes committed by active duty soldiers and veterans (Drummond 2012). In a related essay, Morrison and Casper (2011) explore gendered dynamics of TBI, raising questions about the relative cultural invisibility of violence-induced TBI among women who are abused vis-à-vis sports- and military-related TBI.

Traumatic brain injury can be understood as "a constellation of bodies, knowledge, statistics, diagnoses, psychiatric technologies, cultural meanings, and popular depictions" (Morrison and Casper 2012). Our understandings of TBI are also shaped by economic factors, such as the profits accruing to professional football leagues; by geopolitical factors, such as ongoing military conflicts; and by structural factors, such as gender inequality that leads to domestic violence and differences in power that lead to child abuse.

Mental illness or neurodiversity?

The CNS cannot be discussed without a consideration of mental health and illness. After all, we typically locate mental disorders inside the brain, whether we are discussing schizophrenia, depression, anxiety, or kleptomania (the compulsion to take things)—although we also locate these disorders in the genes, suggesting a hereditary basis to many conditions (Hennah and others 2006). Increasingly, scientific research focuses on neurological aspects of mental health and illness.

There are a wide variety of known disorders that affect mood, cognitive ability, personality, emotions, sleep, sexual function, development, and other aspects of human life. Understandings of these aspects of human experience vary widely across cultures (Kleinman 1987). Moreover, established "scientific" findings are presented in the continually updated "bible" of psychology: the DSM, or Diagnostic and Statistical Manual of Mental Disorders. The DSM guides diagnosis, clinical practice, and treatment protocols, such as pharmaceutical use.

Sociologically, however, what constitutes a "disorder" is understood as a social process, a way of making sense of people's experiences in the world clinically and institutionally. Diseases may have underlying biological components, but their definition, classification, treatment, and meanings *as* diseases and disorders are shaped socially, culturally, and politically. The DSM itself is a social fact, remade with each iteration to include new "disorders" while excluding aspects of life no longer considered "diseases," such as homosexuality (Silverstein 2009). Sociologist Anne Figert (1996) offers an excellent account of the making of premenstrual syndrome (PMS) as a disease.

The *medicalization* of mental experiences and conditions means that many of the "disorders" included in the DSM have become interwoven with the pharmaceutical-industrial complex, a term used by many to describe the for-profit drug industry. Increasingly, where drugs are available, people are targeted for therapeutic drug interventions rather than (or in addition to) other kinds of interventions. From SSRIs (selective serotonin reuptake inhibitors) for depression and anxiety, to fluvoxamine for obsessive-compulsive disorder, to a range of anti-psychotics, people—especially those in the Global North—are heavily medicated.

Anthropologist Joe Dumit (2012) describes this reliance on pharmaceuticals as "drugs for life." These drugs rely on the brain's plasticity and its ability to respond to chemical interventions. Dumit suggests that clinical trials have been outsourced to drug companies, whose motives are profit rather than health. Use of drugs has created new subjectivities, by which many people understand themselves as inherently sick and in need of intervention. However, while drugs may be widely prescribed in the Global North, in many parts of the world life-saving drugs for HIV/AIDS and other diseases are either unavailable or only available as part of clinical trials in which already vulnerable people may be further compromised (Cohen 2006).

Counter to medicalized understandings of mental health and illness, some disability activists and providers have begun to use the terms neurodiversity and neurodivergent. *Neurodiversity* refers to the idea that diverse neurological conditions and experiences are not diseases, but rather normal variations in the human genome. The term challenges scientific and biomedical understandings, such as those reflected in the DSM, as limiting and potentially harmful (Fenton and Krahn 2007). *Neurodivergent* is often used interchangeably with neurodiversity. The use of these terms by activists, health care providers, and others is meant to promote neuro-equality and neuro-inclusivity. The terms suggest a profound shift in how we conceptualize, treat, and respond to people with mental variations, a shift that could ultimately reduce stigma and improve care and support for neurodivergent people.

A CLOSER LOOK

Mad Pride

Mad Pride is a social movement that emerged at the close of the 20th century, with roots in earlier writings and protests focused on unethical treatment of people designated as mentally ill, including forced institutionalization. The heart of the movement, which is dispersed geographically, is obliteration of the stigma attached to mental illness (Glaser 2008). Akin to the disability rights movement, Mad Pride activists take issue with negative cultural representations of madness as "craziness" and people with mental differences as "nutters" and "crazies."

Fully embracing concepts such as neurodiversity, Mad Pride activists reclaim madness for their own purposes, including liberation (Curtis, Deller, and Esther 1999) and fostering public education about various conditions. Although there is no one version of Mad Pride, organizations tend to advocate for more humane mental health services or for no mental health services at all, as well as for accountability from providers and pharmaceutical companies. They also work to counter discrimination and to show that neurodiverse people can lead productive lives.

Some key organizations involved in Mad Pride include MindFreedom International, which works for human rights in mental health services; the National Empowerment Center in the U.S., whose mission is recovery, hope, healing, and empowerment for people with mental health issues; the World Network of Users and Survivors of Psychiatry (WNUSP), which advocates for human rights for users and survivors; and the Icarus Project, which works to create new language and culture that resonates with people's actual experiences. The project's tag line is "navigating the space between brilliance and madness."

All these organizations, and the Mad Pride movement writ large, exemplify the ways that embodied experiences of mental health and illness translate into social activism and the quest for justice. They also illustrate that "mental illness" is a contingent label, one subject to medicalization, institutionalization, cultural meanings, power dynamics, and structural inequalities. Within Mad Pride, embodied psychological experiences are not "deviant," but rather a fact of life in broader social, cultural, economic, and historical contexts.

Conclusion

This chapter has explored the central nervous system, including the brain and spinal cord. Our personalities, reflexes, moods, sexual orientations, intellectual abilities, fears, hopes, and desires have been scientifically and often culturally understood to be lodged in the brain. We tend to explain behaviors, actions, and moods as being rooted in neural pathways or synaptic functioning, and problems as due to malfunctions in these same pathways. As this chapter shows, the neuro age finds explanatory power in the central nervous system and its component parts, shaping how we think about, and respond to, human behavior and social action.

See also

Genes; Cells and tissues; Organs; Respiration; Digestion; Reproduction;

Key terms

Neuroplasticity, neurons, synapses, artificial intelligence, neuroscience, neuroethics, scientific racism, medicalization, neurodiversity, neurodivergent

Discussion questions

1 Explaining human behaviors and differences in terms of the brain is not unique to this time period. Why do you think humans continue to describe so much in terms of this organ? What does the brain symbolize? What do explanations lodged in the brain

mean for other ways of explaining differences and similarities between humans?

2 Artificial intelligence challenges us to understand what makes us truly "human." The notion that a computer, a smartphone, or a tablet device can be given the ability to "learn" through functions like "auto-correct" is often treated as a modern convenience. What are some larger social fears or anxieties invoked by artificial intelligence?
3 The experience of traumatic brain injury often occurs in the pursuit of highly masculinized social practices—military service, athletic achievement, risky car driving. In what ways does masculinity offer a particular type of exposure to activities that might lead to impaired brain function? What types of intervention might you recommend to reduce the rate of TBI? What resistance might you anticipate?
4 Mad Pride is an eclectic social movement of individuals who work to destigmatize neurodiversity within human communities. This form of consciousness-raising enables us to reflect upon the deeply pejorative ways we discipline the "crazy," "insane," or "nutty." Do you think this is an effective way to create social change? Why or why not?
5 What do you think is the future for neuroscience? Forecast what might be in store for our central nervous systems in conjunction with innovations in biotechnology and information technologies.

chapter 8

Circulation

Introduction

In physiological terms, the circulatory system is responsible for moving blood and lymph throughout the body, distributing nutrients, oxygen, carbon dioxide, hormones, and cells. Circulation keeps us alive. When circulation stops, such as during a heart attack or exsanguination (draining of blood), human beings can perish. Approximately 2000 gallons of blood is moved through the body daily.

Vital to circulation is the cardiovascular system, which involves the transport of oxygenated blood through the arteries to the capillaries, where it nourishes the tissues, and the return of oxygen-depleted blood through the veins to the heart, where the cycle is renewed. Circulation also requires the pulmonary system, thus involving a variety of body parts including veins, arteries, heart, lungs, and the lymphatic system.

Knowledge of the circulatory system is ancient. The connection between the heart and the arteries was discovered as early as the 16th century BCE, by the Ancient Egyptians, who recorded details in the Ebers Papyrus (Saba et al. 2006). In the 2nd century AD, the Greek physician Galen theorized arterial function and differential function of the valves. Ibn al-Nafis, an Arabian physician, is considered the father of circulatory physiology for being the first to describe pulmonary circulation, in 1242 (West 2008).

English physician William Harvey was reportedly the first to describe accurately the system of circulation, in the 1620s. Basing his studies in part on human experimentation (e.g., *vivisection*), he detailed the ways blood moves through capillaries and also is transported between the heart and brain through continual pumping (Power 2005). He is sometimes credited as discovering these processes, despite earlier histories of knowledge production, in part because he used the word "circulation" to name the system of blood flow.

But why was circulation called circulation?

Meanings of circulation

The term "circulation" stems from the mid-15th century and derives from the Middle French *circulation* and the Latin *circulationem*, which comes from *circulare*, which means to form a circle, which itself comes from *circulus*, or small ring. Thus, circulation shares a root with circle, and both refer to the notion of a ring. It makes sense that William Harvey, who understood circulation to be a closed system—that is, with blood remaining inside the circulatory system—would recognize the process as a kind of circle or ring.

Circulation is also meaningful in the context of publishing, which in the 1450s offered the first book ever printed with moveable type, the Gutenberg Bible. Subsequently, books, newspapers, magazines, and pamphlets were circulated for public (and private) consumption, and some of these came to be known as circulars. Printed ideas and knowledge thus circulated the world at a more rapid pace than previously in history.

In the world of publishing, circulation is a measure of output and success, with the number of issues distributed revealing interest, support, and relevance. The rise of the Internet and social media means that ideas can circulate much more quickly now, in seconds, leading to even more rapid dissemination of ideas.

We also often use the term "circulating" to describe movement, either across a room (such as a party guest who "circulates" among others socially), or globally, as in "ideas circulate around the world." The planet's circular form lends itself nicely to metaphors of circulation, as for example the idea that maritime explorers circumnavigated the entire globe.

These notions of circulation are not separate from the words we use to describe the human body, and indeed, the naming of the circulatory system reflects the social location, episteme, and linguistic and intellectual privilege of those who "discovered" and named it.

The heart of the matter

The heart is both an organ and a muscle, beating about 100,000 times per day. It appears in this chapter—rather than with "Bones and muscles" or "Organs"—because its role in circulation is so unique and its cultural meanings so significant. The heart is often considered the "center" of circulation, and of the human body itself. Ancient Egyptians believed the heart was the seat of wisdom and the soul (Strudwick 2006). Aristotle, writing in the 4th century BCE, located intelligence and sensation in the heart.

The heart, about the size of an average adult's fist, is an involuntary muscular organ that has been described as resembling, for example, an upside-down pear. Divided into the left and right atria and the left and right ventricles, the heart moves deoxygenated and oxygenated blood through the vascular system. Hearts "beat" as part of the cardiac cycle, in which the valves open and close. Blood pressure refers to the flow of blood from one beat to the next, and the heart rate is the cycle's frequency.

The heart has long been identified as the locus and source of love and our capacity to feel intense emotion toward another. Connections are frequently made between the "life-giving" qualities of love and affection and the vital function of circulation. Indeed, love—whether romantic, sexual, filial, or platonic—is represented in many cultures with a symbol of a heart, its distinctive curved edges meant to evoke the shape of actual hearts.

For example, a common Valentine's Day symbol used throughout Western cultures is a heart rent through with an arrow, presumably shot by Cupid, a figure deriving from the Greek god Eros (Grafton, Most, and Settis 2010).

When love is lost, or we lose a loved one through death or separation, we speak in some cultures of a broken heart. Religious and popular literature from the Bible through Shakespeare through contemporary Western self-help literature is full of references to broken hearts. In a compelling example of the ways that mind and body connect through social life, people who suffer the loss of a loved one and ultimately die from heartbreak are said to suffer from *broken heart syndrome*, also known as stress cardiomyopathy (Akashi et al. 2010).

A rapidly beating heart can be caused by physical exertion, anxiety and stress, fear, or illness. We sometimes say, "My heart was beating so fast!" to demonstrate our degree of terror, which upon recounting can attain a rather pleasurable quality. For example, after a rollercoaster ride or a horror film, you might describe your beating heart to friends in a show of mutual shared survival.

In addition to love and wisdom, the heart has also been considered a moral center of human existence, along with the brain with its presumed rationality and logic. Sometimes in pledging fealty, a knight or soldier might place his hand across his heart, essentially promising his life. In some cultures, when promises are made between individuals, one might speak of crossing one's heart. In the U.S., children sometimes say "cross my heart and hope to die" to indicate the seriousness and honesty of their intent.

The English phrase "to eat one's heart out" derives from the 16th century, when it was used to connote the idea of suffering in silence. A

FIGURE 8.1 Valentine's heart
© Shutterstock

person consumed with grief, rage, bitterness, envy, or despair might be said to "eat one's own heart." Here, the heart is equated with the essence of being or life. It can also refer to an obsessive longing for something, or can be tossed at somebody else as an insult ("Eat your heart out!").

Of course, actually eating a human heart would be cannibalism, and in most settings would be perceived as repugnant. Non-human hearts, such as those from fowl such as turkey, are a different matter, and many humans find these to be tasty treats.

The managed heart

Sociologist Arlie Russell Hochschild coined the term "the managed heart" in her 1983 book of the same name. Used widely in the classroom and translated into German, Chinese, and Japanese, the book offered one of the first sociological investigations of emotion as a social production rather than rooted in instinct or a biological process.

Investigating flight attendants and bill collectors, Hochschild explored the management of emotions and feelings in commercial settings, arguing that people learn what emotions to display in certain contexts. There are, she demonstrated, "feeling rules" that structure human emotion. Her work explores how it is we know how to respond emotionally in the workplace, such as expressing warmth, friendliness, or hospitality through a smile or kind tone in our voice. Convincingly expressing these emotions in line with the "feeling rules" at our jobs may lead to greater job security or token financial reward (e.g., better tips).

However, Hochschild significantly showed that managing emotions has costs, such as a sense of estrangement between what a person "really" feels and the emotion they're allowed or compelled to display. *The Managed Heart* spurred a new subfield in sociology focused on emotions, and presaged contemporary interest in affect studies. In Hochschild's titular framing, managing emotions is akin to managing one's heart.

Politics of blood

Blood is also a crucial component in circulation; indeed, along with lymph it is precious "red gold" (PBS 2002) that sustains human life. Not only is blood the elixir of human life, without which we would perish, the substance is also metaphorically and historically rich.

Blood circulates through and animates our bodies *and* our social lives; indeed, it is essential to what (and who) counts as a human being. For example, blood figures prominently in ideas about contagion, especially since the emergence of HIV/AIDS in the 1980s, and in the long history of diseases such as malaria and sickle cell anemia (Wald 2008), discussed earlier.

Aphorisms such as "blood is thicker than water" signal the importance of biology, family, genealogy, and genetics. Blood is made to stand in for race and ethnicity, and has a particularly dubious history in the context of eugenics (Polsky 2002). It is also deeply gendered, for example in relation to women's menstrual cycles and cultural beliefs about purity and pollution (Stein and Kim 2009).

Blood is a vital ingredient and resource in health care, often heavily commercialized as in blood banks and biomedical research (Carney 2011). Blood is represented in popular culture in manifold ways, from fascination with vampires to a love of forensic technology to the popularity of "slasher" and other violent films.

Thus, blood is simultaneously a substance, object, commodity, signifier, resource, fetish, component, and plot device.

A CLOSER LOOK

Exchanging blood

In many settings, people donate blood for use by other people of compatible blood types. In the United States, the American Red Cross supplies 40 percent of the nation's blood supply for use in medical emergencies, disasters, and by hemophiliacs, with some four million people donating. The National Blood Service, part of NHS Blood and Transit, serves England and Wales. Several other countries have similar voluntary blood donation systems and organizations.

However, in other settings blood is commodified, with some people selling their blood to others through various networks. Most often, people who are economically disadvantaged are the sellers. In addition to reflecting and perpetuating global inequality, marketized blood economies may exacerbate health problems. For example, in the 1990s China's "Blood Plasma Economy" campaign, in which poor peasants exchanged their blood for money, led to high HIV infection rates in recipients of blood (Rogowska-Szadkowska 2011).

Fears of HIV infection and a "tainted" blood supply led to technical and institutional changes in blood banking. In the early days of the HIV/AIDS crisis, the mechanisms of transmission were not fully understood, and many people were infected via blood transfusion. Some countries, such as Canada, were particularly hard-hit, with their entire supplies affected in ways that still resonate (Lanteigne 2013). Today, questionnaires and screening procedures, as well as the ability to detect the virus, has reduced the risk of infection via transfusion to approximately one in 1.5 million, according to the U.S. Centers for Disease Control and Prevention.

There have also been policy measures introduced, some controversial. The U.S. Food and Drug Administration bans blood donations from men who have had sex with men (MSM), even one time, since 1977. (Men in this category can donate blood for their own health, but not for use by others.) The ban has been the subject of bioethical and legal debate, as many view it as discriminatory (Galarneau 2010). In 2011, a similar ban in the UK was modified to prevent MSM from donating if they had been sexually active in the previous year (Gallagher 2011).

However, HIV is not the only pathogen that can threaten the blood supply. In 1999, the United States banned donations from people who had visited or lived in Britain or Europe between 1980 and 1997, out of fear that the blood supply could become contaminated by acquired Creutzfeldt-Jakob or "mad cow" disease (O'Neill 2003). Several countries restrict donations from people who have been outside the country for an extended period of time. Australia restricts blood donations within six months after receiving a tattoo.

There is no international regulatory body for blood donation, although the World Health Organization provides technical and regulatory assistance to national authorities. Different countries have different systems and rules, making for a sometimes-confusing patchwork, especially when blood travels between countries legally or illicitly, as in the "red market" (Carney 2011). The co-existence of both unpaid and paid exchanges further complicates matters, as different ethical and legal issues pertain.

Cardiovascular health and disease

The World Health Organization estimates that 17.3 million people died from cardiovascular disease (CVD) in 2008, with low- and middle-income countries disproportionately affected. The two leading causes of death in the world are ischemic heart disease (11.2 percent of deaths) and stroke (10.6 percent of deaths). Both are cardiovascular—that is, related to the circulatory system. Each of these diseases carries with it various meanings about "lifestyle" contributions to morbidity and mortality.

Ischemic heart disease is also known as coronary artery disease or atherosclerotic heart disease. Plaque accumulates along the inner walls of the arteries, which restricts them and decreases the flow of blood to the heart. The disease may progress silently as plaque accrues, and when blood flow becomes too restricted, BAM! A heart attack. Cardiologists are the medical specialists tasked with diagnosing, preventing, and treating heart disease, including heart attacks.

Risk factors for ischemic heart disease have been identified, including hypertension (high blood pressure), high cholesterol, smoking, hyperglycemia (high blood glucose), and "Type A" behavior (Type A behavior is described as someone who is rigid, highly organized, and ambitious) (Wilson and others 1998). Lack of physical activity is also sometimes included as a risk factor, and is often mentioned in discussions of "poor circulation." In short, fatty foods, cigarettes, and not exercising—i.e., so-called lifestyle factors—are believed to contribute substantially to CVD.

Stroke, the second leading cause of death in the world, occurs when blood flow to the brain is interrupted. Here, too, restricted or blocked arteries may be responsible. This kind of stroke, called an ischemic stroke, accounts for about 85 percent of all strokes, and thus like heart attacks, is frequently connected to "lifestyle" factors. Less common are hemorrhagic strokes, caused by leakages or ruptures of blood in the brain. High blood pressure can lead to hemorrhagic stroke.

Although heart attacks and strokes are the leading causes of death around the world, they do not affect everyone equally. Men tend to have more heart attacks and strokes, and often at earlier ages than women. However, women are more likely to die from their symptoms, in part because heart disease has long been considered a male disease and women are frequently underdiagnosed or improperly diagnosed (Wenger 2002). Indeed, several major heart studies excluded women (Wenger, Speroff, and Packard 1993; Holdcroft 2007).

Beyond these gender differences, there are other differences, too, for example in access to diagnosis and treatment in different parts of the world. Socioeconomic status (SES) is related to cardiovascular disease, with people who earn less money consistently showing a higher disease incidence. Low SES is also correlated to higher rates of smoking and obesity and to lower rates of education, according to the World Health Organization.

While "lifestyle" is deemed the culprit in CVD, including heart attacks and stroke, socioeconomic status complicates our understanding. Certain "lifestyle" risks, such as smoking and nutritional intake, are related to poverty. Is poverty, then, also a "lifestyle," or are there other, more structural issues at play?

"Poor circulation" and "lifestyle diseases" are implicated in systems of global inequality, and call into question the various ways we measure and make sense of morbidity and mortality. Because heart attack and stroke, along with other diseases such as diabetes, presumably are caused by "sedentary" behavior and bad habits, issues of responsibility and blame become salient. Certain types of bodies—for example, the obese—are seen as both undesirable and pathogenic (Saguy 2013).

The subfield of fat studies has interrogated these kinds of constructions and their relationship to medical knowledge and practice, raising important questions about health, embodiment, and science. Countering oppression of fat people, the size acceptance movement emerged out of feminist and queer liberation movements. Feminists have been advocating on behalf of fat bodies for decades (Kent 2001). As Solovay and Rothblum (2009: 3) write: "The assumption that fat people are unhealthy is so ingrained in western society that it is hard to get people to face the facts."

Acknowledging that fat bodies are not necessarily unhealthy—or "grossly" and "pathologically" obese, as they are often termed in the popular media—would also help in the acceptance of larger and different bodies. A grassroots movement of healthcare workers, researchers, consumers, and activists has been working to shift the idea of what a healthy body looks like. Health at Every Size (HAES) emphasizes healthy everyday practices and body acceptance. The actual weight and size of a body is not as important as its overall health (Burgard 2009; Lupton 2013).

A CLOSER LOOK

Non-human hearts

Sometimes, heart failure can be "treated" (but not cured) by replacing the damaged heart with the healthy organ of a deceased donor. This typically happens during the "end stage" of heart failure, when the heart cannot pump enough blood to meet the body's needs and other treatments have failed.

Heart transplantation (also called cardiac transplantation) is relatively rare, with only about 3,500 transplants worldwide. About two-thirds of these are in the United States, which has the world's most expensive health care system. Of course, in settings where treatment or transplantation is not available, heart failure leads to death.

Human heart transplants can extend both the quantity (years lived) and quality of life, but cost a great deal of money. (Approximately US$997,000 for a heart transplant, and US$1.15 million for a heart-lung transplant.) Because the number of people with heart disease exceeds the number of donor hearts available, research has focused on alternatives to human heart transplants. In the U.S. alone, some 5.7 million people experience heart failure each year, and in the UK, 8,000 people are on the transplant waiting list for new organs.

Non-human animals have long been investigated as potential sources of healthy donor hearts, through a procedure known as a *xenograft*. In 1984 in California, for example, Stephanie Fae Beauclair, or "Baby Fae" as she was known publicly, was fitted with a baboon heart after she was born with hypoplastic left heart syndrome. She died 21 days later. The case provoked ethical concerns about the use of animals, informed consent in children, and experimental medicine (Kushner and Belliotti 1985).

Animal to human transplants of organs, tissues, and cells poses difficulties, including rejection of the donor parts by the recipient body. That is, the recipient's immune system attacks the foreign nonhuman donor tissue. Also, the use of animals without their consent to enhance human health poses major ethical dilemmas (Hughes

1998). For example, do humans have a right to use animals, and to interfere with nature in these ways? Use of animals also raises the specter of animal diseases being introduced into human populations.

Given the perceived need for available hearts juxtaposed with a limited supply, and also taking into account the ethical and practical concerns related to nonhuman animal transplants, scientists have turned to two other cutting-edge possibilities: stem cells and prosthetics.

In cardiac stem cell therapy, the stem cells are introduced into the body to repair the damaged valves, tissues, and muscles, specifically the myocardium (Oettgen 2006). Both embryonic stem (ES) cells and adult stem cells have been used, as have a patient's own cardiac stem cells. The cells can be infused directly into a patient's coronary arteries, or injected into the ventricular wall. Several clinical studies have demonstrated that stem cells may help to promote tissue regeneration in hearts, and thus ultimately prevent heart failure, although the procedure remains experimental (Zhang and Pasumarthi 2008).

Clinical scientists and medical engineers have also pursued development and use of an artificial or prosthetic heart. Several innovations appeared from the mid-20th century on, including heart-lung machines, artificial heart valves, and eventually a fully artificial heart comprised of materials such as plastic and titanium. The first patient to receive an artificial heart for treatment of heart failure was Barney Clark, a retired dentist, in 1982 in Salt Lake City (Jauhar 2004).

Patients are often fitted with artificial heart valves as a "bridge" to human heart transplantation. Others are provided with artificial hearts on a more permanent basis. Currently, only one artificial heart, made by SynCardia in Tucson, Arizona, has regulatory approval in the U.S., Canada, and Europe, and thus dominates the market. The company's Total Artificial Heart replaces the ventricles and all four heart valves.

In 2013, researchers announced a new *bioprosthetic* heart made of both biological and synthetic materials (Young 2013). The collaboration between Alain Carpentier, a cardiac surgeon who innovated heart valve repair, and the European Aerospace and Defense Systems will be tested in the Middle East and Europe. The researchers hope the hybrid design will help prevent issues of immune rejection.

In 1919, in *The Devolutionist and the Emancipatrix*, science fiction writer Homer Eon Flint described a patient being fitted with an artificial heart. Now, almost 100 years later, the speculative has become reality.

Conclusion

The circulatory system is a super highway inside the body, transporting nutrients, water, and oxygen as well as waste and carbon dioxide. This intricate system connects all the cells of our body and enrolls multiple body parts in doing so. We can describe in standard anatomical or biological terms the functions of the heart, blood, and blood vessels comprising the circulatory system. However, as we have shown, these body parts and systems are imbued with social and cultural meanings and also, in turn, inform our everyday lives.

From sharing matters of the heart with our confidantes and love objects to developing universal precautions for the management and containment of blood, our daily lives in many parts of the world are structured by emotional and practical engagement with the circulatory system. Leading causes of death globally are attributed to dysfunction in this system, so much so that we innovate ways to improve or replace the mechanisms of circulation through synthetic or nonhuman animal organs.

Whether these innovations will help the people most in need—those dying of heart failure in low- and middle-income countries, remains to be seen. As we have discussed throughout this book, health and illness are stratified, as are treatments and cures.

See also

Cells and tissues; Organs; Fluids; Bones and muscles; Respiration

Key terms

Vivisection, episteme, broken heart syndrome, xenograft

Discussion questions

1 The heart is often considered the center of the body and the seat of honesty, wisdom, and love. However, the 21st century has also been termed the "neuro" age, with an emphasis on the brain, as we discussed in the previous chapter. Which of these—heart or brain—do you think is the best symbol of humanity, and why?
2 Research has shown that it may indeed be possible to die of a broken heart. Given the demonstrated role of grief and stress in

cardiomyopathy, what are some nonmedical ways that a "broken heart" might be treated?

3 What type of emotion work is required of you in your day-to-day experiences? Where can you be your most authentic self? When are you "faking it"? How do you know the difference? Is it ever confusing or difficult to figure out what is the "real" you or the one performing for others?
4 What are some of the ways you see blood symbolized in everyday life? Are there images you come across that signify blood?
5 Discuss some of the ethical considerations regarding heart transplantation. Are there different social and cultural consequences of a human heart versus a synthetic heart versus an animal organ?

chapter 9

Respiration

Introduction

Humans and other earth-bound organisms are surrounded by a colorless, odorless, tasteless, gaseous mixture, which contains mainly nitrogen (approximately 78 percent) and oxygen (approximately 21 percent), and lesser amounts of argon, carbon dioxide, hydrogen, neon, helium, and other gases. This atmospheric gas, or air, is part of what makes our planet habitable by flora and fauna, including us.

In short, we must breathe to survive.

An adult's respiratory rate, or the number of inhalations and exhalations taken in a 60-second period, is between 12 and 18 breaths. During ventilation, human bodies exchange the gases carbon dioxide and oxygen. This exchange takes place in the alveoli of the lungs. Element transference is necessary for the functioning of our bodies, but there are sometimes experiences when we find it necessary to hold our breath. For most people, a minute or so is the limit of holding the breath. Beyond this, we would lose consciousness owing to a build-up of carbon dioxide in our bodies and oxygen deprivation, which can lead to brain damage.

Lung capacity and the ability to hold one's breath varies among human beings, with some people able to cease their breathing for longer periods and others (especially those with respiratory conditions) unable to hold their breath for any length of time. In May 2012, for example, world-record-holder and champion free diver 39-year old Stig Severinsen held his breath for 22 minutes and now trains people in how to control breath (Severinsen 2010).

In this chapter, we explore the human respiratory system in the context of larger ecosystems in which we live. Damage to the vital resource of air has occurred across centuries, negatively affecting human health. An increase in environmental toxins such as lead has given rise to environmental justice movements. We examine in particular the increase in asthma rates. We also show that information is revealed through our breath, leading to the innovation of breath analysis as a system of monitoring and surveilling human behavior. End-of-life decision-making also invokes breathing, as when mechanical ventilation is in use.

We offer a closer look at the role of gas masks in human history and how they protect our ability to breathe while we are simultaneously attempting through militarization to eliminate breath (and life) from others. Another "closer look" explores the popularization of the practice of yoga.

Baby's first breath

Prior to birth, a fetus is dependent on placental oxygen delivered through the umbilical cord using the pregnant woman's circulatory system. In most births, an infant's lungs are full of fetal lung fluid, but after about 10 seconds a baby takes its first breath. The fluid in the lungs is reabsorbed into the baby's blood and lymph system. A loud and lusty cry announces that the baby is able to take in air through its nose and mouth and through its airway into the lungs.

The baby, often purplish at birth, quickly turns a pinker color once it intakes air, even with babies whose skin becomes more pigmented as they grow. Sometimes to encourage babies to breathe, medical personnel will suction fluid from the baby's nose and mouth to clear the newborn of all secretions. An infant's autonomous breathing is further achieved through cutting the umbilical cord; the lungs inflate and begin to work on their own. Newborns make many different sounds when breathing—gurgling, snorting, whistling, and hiccuping.

Some babies have trouble breathing at birth. Hypoxia, or the deprivation of adequate oxygen supply, is caused by a variety of circumstances including the umbilical cord wrapped around a newborn's neck at delivery. Complications of hypoxia include learning disabilities, cerebral palsy, fetal brain injury, or even death. The World Health Organization reports that four to nine million newborns experience hypoxia annually, leading to approximately 1.2 million deaths. Hypoxia is a leading cause of sudden infant death syndrome, or SIDS (Saugstad and Rognum 2003), which is the third leading cause of death of newborns worldwide.

Air pollution, climate change, and the global burden of disease

Respiratory infections, cardiovascular health, lung cancer, and asthma have all been linked to increased levels of air pollution worldwide. In fact, as reported by the United Nations, both indoor and outdoor air pollution kills more people every year than AIDS and malaria combined (Richard 2013).

The World Health Organization defines *air pollution* as the contamination of indoor and outdoor environments by any chemical, physical, or biological agent that modifies the natural characteristics of the atmosphere. *Particulate matter* (PM), which makes up air pollution, is a complex mixture of extremely small particles and liquid droplets including smoke, dirt, dust, mold, spores, pollen, toxic organic compounds, and heavy metals. Exposure to air pollution has been linked to stroke, pneumonia and COPD (Chronic Obstructive Pulmonary Disease), heart disease, cystic fibrosis, asthma, chronic bronchitis, emphysema, and multiple forms of cancer.

Industrial facilities, forest fires, motor vehicles, and household combustible devices (burning wood, animal dung, crop waste, or coal) are the most common contributors to air pollution. Household combustible devices create indoor air pollution, exposing mostly women and children in the domestic sphere as they are primarily responsible for cooking food.

According to the International Energy Agency (IEA), people who use biomass fuels for household purposes will increase from today's 2.4 billion

FIGURE 9.1 India pollution
© Prakash Singh/AFP/Getty Indian commuters wait for a bus early on a polluted morning in New Delhi. This is an example of smog, the combination of fog and pollutants such as smoke or sulfur dioxide. (Photo: Prakash Singh/AFP/Getty Images)

globally to 2.7 billion by 2030. This accounts for 90 percent of household energy consumption in the Global South by approximately one-third of the global population.

Climate change refers to long-term changes in the earth's temperature, precipitation, wind patterns, sea level, and snow cover. Significant controversy surrounds the anthropogenic (or "man-made") contributions to climate change. A comprehensive analysis of scientific literature covering two decades has reached conclusive consensus that climate change is mostly caused by human activities. Energy extraction and use have produced the lion's share of man-made greenhouse gases – especially CO_2 – that cause unwanted climate change.

Greenhouse gases emitted through the burning of fossil fuels are one of the primary contributors to climate change. The human body, and in particular the respiratory system, is deeply affected by our species' pursuits in transportation and other industries. Climate change has increased morbidity and mortality in many regions of the world (Patz et al. 2005), and specifically elevated rates of respiratory illness owing to heat waves and corresponding cardiovascular mortality.

The global burden of disease is not equally distributed and those who contribute the least to processes of global warming are at the greatest risk of being negatively affected by climate change. A team of scientists working for the Global Health Commission (Lancet Commissions 2009: 1694) stated:

> Climate change will have its greatest effect on those who have the least access to the world's resources and who have contributed least to its cause. Without mitigation and adaptation, it will increase health inequity especially through negative effects on the social determinants of health in the poorest communities.

For example, there is a global rise in asthma prevalence. *Asthma* is a chronic lung disease that inflames and narrows the airways. The prevalence of asthma has increased *50 percent every decade* (Braman 2006), and asthma is the most common chronic disease among children. There is a positive correlation between climate change, which worsens conditions of air pollution and alters the pollen season, and increased rates of asthma (Amato 2013).

Sociologist Andrew Szasz (2007) has engaged in critical examination of social practices and popular discourses surrounding possible solutions to air pollution. His work has demonstrated a reliance on private or individual solutions to ecological and health crises brought on by air pollution rather than pursuing social structural changes. In particular, Szasz (2007: 165) investigates how higher status Americans who are aware of environmental degradations, such as air pollution and toxins in the food supply, do not push for regulatory controls over commercial enterprises, but rather believe they can purchase themselves out of danger. Individuals buy organic produce or invest in personal air filters:

> Disposable income is the key to who can make these consumer choices and who cannot. You can reduce exposure to outdoor air pollution only by spending hundreds of thousands of dollars or more for a home away from the industrial zone and the traffic corridor. In every home, every organic alternative costs more—and sometimes a great deal more—than its "conventional" counterpart.

According to Szasz and other scholars, individual consumption practices do not modify the broader environmental threats and actually cannot lead to the necessary reforms to create healthier bodies and environments.

Surgical masks

At the end of the 19th century, hospitals began to introduce surgical masks among medical practitioners to prevent the spread of infection, and common use increased during the global Spanish 'flu pandemic of 1918 (Spooner 1967). The surgical mask is used to protect health care providers and patients during procedures that expose vulnerable parts of the body to possible transmission of germs. These masks are most effective at limiting the wearer from exhaling contaminants and spreading disease to others through respiration, and they also can protect the wearer from inhaling contaminants, such as from patients suffering from contagious illnesses and infection.

Over the past few decades, wearing a surgical mask to limit one's own exposure to germs, pollen, dust, and particulate matter from air pollution is increasingly common in many Asian countries. Studies have demonstrated some potential benefits to wearing a surgical mask and reducing cardiovascular effects of pollution (Langrish et al. 2009).

As the incidence and prevalence rates of influenza strains, such as H1N1, increase along with rates of viral infection, such as severe acute respiratory syndrome (SARS), demand for surgical masks has increased. In fact, during particularly virulent episodes of swine flu in Japan, pharmacies were known to sell out of surgical masks. Wearing of masks has become so popular and integrated into everyday life in some countries that they are available in different styles and designs.

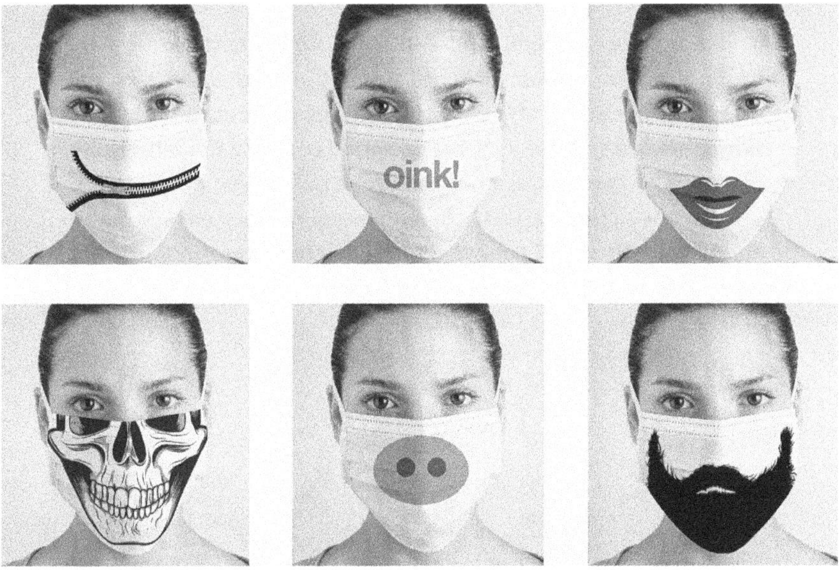

FIGURE 9.2 Pollution masks
Photo courtesy of Irina Blok

A CLOSER LOOK

Work, war, and gas masks

Beyond the more common, low-tech surgical mask, a *gas mask*, also known as a respirator, is a device designed to filter out harmful gases and particles. In the early 1900s, innovative technology for modern gas masks was configured for coal miners, firefighters, and underwater divers. Today, gas masks (and atropine shots) are a staple of military and civilian work at chemical weapons armories and disposal facilities around the world (Casper and Moore 2011). Gas masks range from a half-mask to full-sized air-purifying respirators to the SCBA (self-contained breathing apparatus) system.

Because our respiratory system is one of potential great vulnerability, it can be exploited during military conflicts through aerosol attacks inflicting mass casualties. Hence, gas masks are also deeply connected to the military industrial complex and the rise of chemical warfare. In many countries in Europe, gas masks were distributed to the civilian population during World War II. Studying the role of gas masks in children's material culture, archaeologist Gabriel Moshenska (2010: 615) explains:

Unlike bullets, shrapnel, and explosives which destroy the vital organs, poison gases incapacitate and kill their victims by attacking the seats of the senses: the eyes, nose, mouth, and skin. Gas masks, conversely, are designed to shield the sensorium, the seats of the senses, from the effects of poison gases. In the process they create a restrictive, minimal sensory environment: an anaesthetic shield against an anti-aesthetic weapon.

FIGURE 9.3 Gas masks 1940s.
© Muffinn/Flickr
"Gas mask practice Hallow School 1940s on Hallow Village Green – Jean Woodward"

Following the trend from World War II, a nation's military usually supplies its soldiers (and sometimes military animals, such as horses and dogs) with gas masks. Additionally, at times of increased perceived threat of chemical attacks, countries such as Israel have set up distribution centers for gas masks.

During internal civil conflict, individuals may attempt to fashion their own gas masks from readily available materials and instructions on the Internet. For example, in Syria fighters in the Free Syrian Army are creating do-it-yourself gas masks. As Syrian Abu-Tarek stated: "I use information that I found on the internet and from my military service in order to create masks made from an empty container, coal and cotton soaked in a carbonated liquid drink" (Okabi 2013).

Social class and lung health

As discussed in previous chapters, lungs are two spongy organs in the chest cavity that take on the chief function of inhaling and exhaling through work of a muscular partition called the diaphragm. In conjunction with the heart, the lungs circulate oxygen throughout the body. Since the human body is embedded in both the physical environment *and* social systems, the functioning of bodily systems is conditioned and constrained by unequal distribution of threats to optimal functioning. In the case of respiration, lung health is linked to social class and occupation.

For example, one study followed a socioeconomically representative cohort of 7735 British men aged 40–59 years for a 20-year period to determine the association of their lung health with their socio-economic position (Ramsay et al. 2011). The socioeconomic position of the adult men was derived from longest-held occupation in middle age and their childhood socioeconomic position from their father's longest-held occupation. Even when the researchers adjusted for factors of age, cigarette smoking, body mass index, physical activity, and history of bronchitis, men who ranked lower on SES scales had poorer lung health. In fact, men in manual social classes, often referred to as "blue collar" workers, in both childhood and adulthood had the greatest decline in lung function compared to those in nonmanual groups, or "white collar" workers.

Turning to another example of social location enmeshed with respiratory health, sociologist Barbara Ellen Smith (1987) studied the illness known as "black lung disease," a preventable occupational disease. Coal dust is one of the main causes of lung ailments, such as coal workers' pneumoconiosis (CWP), emphysema, silicosis, and bronchitis—conditions known collectively as black lung. The disease leads to lung impairment, permanent disability, and premature death. Smith's work offers a significant example of differences between lay people and the medical industrial complex in defining disease and its causes.

Black lung disease emerged in West Virginia in 1968, spurring a controversy over who would control definitions of the disease. The political economy of coalmining, as well as the entrenched relationship between the coal industrialists and the "company" doctors, influenced the "medical" diagnosis of black lung disease, as well as understanding of its causes. Company owners and doctors emphasized that it was the miners' lifestyle that led to respiratory illness. The miners and their families mobilized to demand that black lung be recognized as a legitimate disease eligible for workers' compensation. Advocates further organized for federal standards to control working conditions and environmental risks in coalmines.

Examples such as the black lung disease controversy are exemplary of *popular epidemiology*, whereby "laypeople detect and act on environmental hazards and diseases" (Brown 1992: 268). The coalminers and their families were successful in forcing some changes in the industry, although miners' health remains an important concern.

Black lung disease has not been eradicated, and in 2012 it was on the rise in the United States and China. Globally, respiratory health is deeply entwined with urban and rural ecologies and marketplaces. Where we work, play, live, and travel affects our lung health. As the next section shows, even in our leisure activities our breath may reveal certain behaviors to others.

Breath analysis

Did you know that you have your own breath signature? The biochemical information in breath can indicate an individual's health status, enabling possible innovations in personalized medicine (Sinues, Kohler, Zenobi 2013). Although infrequently used (for now), breath has been used to determine information about human bodies. Biological analysis conducted on breath can include detection of asthma, lung cancer, and diabetes.

In addition, scientists established that the BrAC (breath alcohol concentration) level is associated with a person's BAC (blood alcohol concentration). The *breathalyzer* is a technological and noninvasive way to measure organic compounds in exhaled breath—most commonly through mathematical modeling that can be related to blood alcohol level. Earlier generations of the breathalyzer, from about the 1930s, were called the drunkometer, the intoximeter, the alcometer, and the alcosensor. In 1954, Robert Borkenstein designed the breathalyzer, which was marketed by Smith and Wesson and soon became the generic name for all breath alcohol testing.

Different countries have different procedures in place regulating how and under what circumstances law enforcement officials can stop vehicles and require a driver to comply with testing. The procedure using a breathalyzer is called a field sobriety test. Typically, preliminary roadside breath alcohol tests are conducted when an officer reasonably suspects a person is driving under the influence of alcohol or drugs.

Personal use of breathalyzers is also on the rise. The U.S. has seen an almost tenfold jump in personal sales over a five-year period (Sy 2009). Consumers self-report using breath analysis tests to determine if they should drive their cars from parties or bars. In addition, parents have used breathalyzers to check on their children's alcohol consumption.

Mechanical ventilation and end-of-life decision-making

In 1927, the negative pressure ventilator, also known as the iron lung, was considered a medical breakthrough to treat polio, an infectious disease that paralyzes the lungs. The large cylinder of steel contained the human body almost entirely. With only a person's head and neck sticking out the top of the apparatus, the sealed, airtight container used air pressure to imitate the pattern of breathing, manually forcing the lungs to function. During the 1940s and 1950s, iron lungs were in heavy rotation at hospitals treating those suffering from polio (Oshinsky 2006).

As polio was eradicated through public health vaccination programs, positive pressure ventilation replaced the iron lung. A *ventilator* is a machine that involves using a ventilated facemask or an endotracheal tube inserted through the throat and down the windpipe, to maintain an airway for a patient and create adequate oxygenation. Ventilators are used during respiratory failure. They do not treat disease or offer a patient recovery; rather, they are used to keep a patient alive. Often, a patient is placed on a ventilator temporarily while health care providers treat the underlying condition or disease that has compromised lung function.

However, sometimes the patient is placed on a ventilator to prolong life, even though there is no treatment to eradicate the disease. This process is a rich, contested topic in *end-of-life decision-making*, or the process of creating advanced directions for the treatment of your body or giving a loved one the right to do so, when you are no longer able to make decisions for yourself.

Mechanical ventilation can also be used in cases of organ procurement and donation. Individuals' bodies can be kept alive through mechanical ventilation for organ recovery and harvesting (De Lora and Blanco 2013). Here, the intent is not to preserve the life function for future survival of the patient, but rather to utilize the patient's body parts for other patients. As discussed earlier in the book, organ transplantation remains a contested field of practice.

To recap, bioethics is concerned with developments in biological and medical sciences, which may proceed rapidly and in advance of consideration of their social impact. Whether or not to keep an individual on mechanical ventilation is an end-of-life decision that falls within the domain of bioethics. The withdrawal of life-sustaining treatment could be considered either a humane option to let someone die so as to end their suffering, or as the deliberate practice of killing someone. End-of-life issues are especially complex when the patient is young.

What are the ethical implications of keeping an individual alive when there is no hope for a cure? What is the best practice, for whom, at what stage of their life? How do "best practices" change as our technology is innovated beyond our body's capacity to heal itself?

A CLOSER LOOK

Yoga

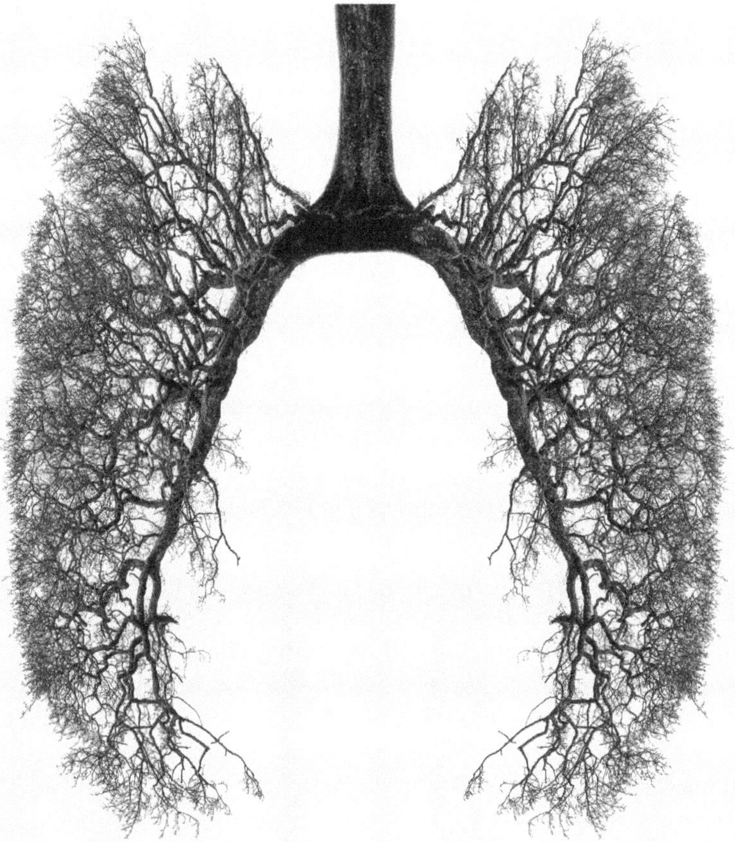

FIGURE 9.4 Lungs
© Getty Images/Mike Hill

One of the most interesting things about the human respiratory system is that while we are alive, we endlessly breathe—without much conscious thought in the matter. We typically become aware of our breathing only when it is labored, for example when we exercise heavily or are in a low-oxygen environment. The chemical exchange of carbon, nitrogen, and oxygen happens consistently when we are awake and sleeping. So subconscious is our breathing that we often do not recognize when we are holding our breath, as we sometimes do during times of stress or fear.

Yoga is an ancient Hindu practice that uses a combination of physical and mental methods in order to reach higher levels of health and inner peace. Breath control is taught through yoga poses and meditation practices. The *prana* is conceived as the vital and original life force, and it is stimulated through training in movement and diaphragmatic breathing. Imported from the Global South, yoga has also become a lifestyle activity in the Global North, where conspicuous consumption of yoga clothing, mats, how-to videos, and vernacular can be found in many major metropolitan areas.

There are a wide variety of yoga practices that stress different aspects of prana and maintain a different flow of poses; these include Bikram, Iyengar, Hatha, Vinyasa, Kundalini, Anusara, and others. Although clearly deeply integrated in Eastern health practices and wellbeing, in the past decade allopathic health researchers have explored how yoga is beneficial for African American heart patients (Pullen et al. 2010), women's smoking cessation (Bock et al. 2012), and those diagnosed with COPD (Fulambarker 2012).

Through a case study of fieldwork in India and France, geographer Anne-Cecile Hoyez (2007: 119) introduces the idea of "globalized therapeutic landscapes" as a way to understand the globalization of yoga. This is the process of cultural, therapeutic, or consumption attitudes taking a practice from one place and bringing it to other locations. She examines how India has become an "emitting pole" of social and cultural practices across the globe:

> From the 1980s to the present day, more and more yoga styles and types appear. One can talk about "indigenous yogas" in the sense that in many countries, such as Brazil or the Czech Republic, yoga was systematised by a small number of persons. Some Brazilians, through readings and through interindividual interactions, have built a local knowledge of yoga (in Sao Paulo or Porto Alegre, for example). A similar situation emerges in Eastern Europe. Just after the fall of communism, a yoga centre was set up in Prague.
>
> Nevertheless, one could argue that this worldwide diffusion is not complete in the sense that some countries are not represented. Indeed, Arab countries, Saharan and sub-Saharan Africa do not appear in world maps of yoga centres. Political and social situations in some of these countries prevent the implantation of yoga centres; or, if there is no formal ban, yoga centres are less numerous, less frequented, and do not generate significant activity.

Yoga has travelled across borders, through and beyond individual practitioners, and evolving conceptions about integrated health and wellness. It has offered new opportunities for individuals to breathe.

Conclusion

While the body's respiratory system may seem entirely internal to the human body, it functions based on continuous exposure to and interaction with the external world. We breathe in the air of our environments and we breathe out not only gaseous waste, but also clues about what we might have imbibed or where we might have worked.

As humans have recognized their respiratory vulnerability to particles and noxious fumes, they have innovated gadgets to protect the lungs. Gas masks and ventilators help us to breathe. Breathalyzers detect things about us through our exhalations, a form of surveillance that can have legal implications.

At the beginning and end of life, some people have access to technologies to maintain their breathing, even when their respiratory system is injured or nonfunctioning. From our first to our last breath, how we breathe and move air through our bodies is contingent on social organization as much as the physical environment. Breathing, like other bodily processes, is stratified, with people in poorer, polluted countries attempting to protect their lungs with surgical masks, and people in wealthier countries debating the ethics of extending lives mechanically.

See also

Organs; Bones and muscles; Circulation; Reproduction; Faces

Key terms

Hypoxia, air pollution, yoga, environmental justice, particulate matter, climate change, asthma, lungs, popular epidemiology, breathalyzer, gas mask, ventilator, bioethics, end-of-life decision-making, yoga

Discussion questions

1. Air is usually invisible, so how does one perceive air pollution? What are some of the ways you may be exposed to air pollution on a daily basis? Have you travelled outside of your local region? What differences do you notice regarding air quality? Are there innovative ways that people in your region and elsewhere manage air pollution?
2. Surgical masks and gas masks are preventative and protective measures that individuals and nation states use to keep people safer. One wonders how much more common use of these devices, and their next generations, will be in everyday life. What types of safety apparatuses can you imagine being used in everyday life? What could you anticipate that would bring on the need for such protection of your respiratory system?
3. Is it an invasion of privacy for someone to use a breathalyzer on another individual? Friends, parents, law enforcement? Why or why not? What are other ways we use the body to reveal evidence, sometimes against the wishes of the person being tested?
4. Prior to the 1900s, when the lungs no longer worked it signaled that death was fast approaching. However, the lungs can now be assisted in order to keep a body alive, sometimes for very long periods. But mechanical ventilation brings about some thorny bioethical issues. Discuss the factors that create a quality of life worth living.
5. Yoga and other practices of exercise and enlightenment associated with the breath are increasingly popular in Western countries. The marketing of slogans like "Breathe" suggest a need for reminders of what the body does anyway. What are some reasons for growth of these practices? To what societal changes might the interest in yoga and "alternative" practices be responding?

chapter 10

Endocrine system and metabolism

Introduction

As humans, we often celebrate our adaptability to different geographies, meteorological conditions, and ideologies. Through human innovations, people have been able to survive in different, sometimes extreme climates, travel across great distances to relocate entire populations, and genetically alter foods for transport and preservation.

But in addition to adaptations facilitated by technological inventions, such as farming, weaving, GMOs, and cryopreservation, there are also physiological adaptations that human bodies manage daily to adjust to changes in our environments.

This chapter reviews metabolic systems. Here, we examine how metaphors of regulation based on our metabolic and hormonal processes are useful to social thinkers to explain social organization, including urban life and ecological degradation.

Overview of metabolism

The *endocrine system* is a collection of glands, including the adrenal, pituitary, hypothalamus, pineal body, pancreas, ovaries and testes, and thymus and parathyroid, which secrete chemical signals called hormones. Hormones are released into the bloodstream and travel to organs and tissues throughout the body. Some of these hormones include melatonin, adrenaline, and growth hormone which signal the body to sleep, produce energy, or develop.

Metaphorically, the bloodstream is like a broadcasting system whereby transmissions of information (hormones) travel throughout the network (the body) and are delivered to different receptors (organs) to produce different effects (cell production or physiological changes in heartbeat). In particular, the endocrine system is responsible for regulating the metabolism of an organism. The thyroid gland is most directly related to metabolism as well as body heat and bone growth.

Metabolism is a term used to describe the sum total of chemical and physical changes that occur in the body when molecules are either broken down or built up in cells and tissues; for example, turning food into energy, repairing damaged cells, or breathing in oxygen and breathing out carbon dioxide.

In biological terms, in humans and other mammals, metabolic properties maintain *homeostasis*, or an inner balance such that despite external or environmental fluctuations, the internal body manages to keep things stable with respect to temperature, blood pressure, and sleep and wakefulness.

Our metabolism also processes and assimilates food and drugs through the processes of *catabolism*, or the breakdown of complex molecules into simpler ones, releasing energy, and *anabolism*, or the building up or growing of new tissues, requiring energy.

A CLOSER LOOK

Gendered hormones

It is now widely accepted that there are "male" and "female" hormones, specifically testosterone and estrogen, which shape human behavior in gendered ways. Yet the history of hormones is immensely revealing, for there was nothing "natural" about their original gendered designation at the dawn of a new field called endocrinology.

Dutch scholar Nelly Oudshoorn (1994), in *Beyond the Natural Body*, traces the "discovery" of these hormones by physician and pharmacologist Ernst Laqueur at the University of Amsterdam in the 1920s. Interested in both the materiality of human bodies and their cultural and scientific representations, Oudshoorn unravels the layered history of the hormones we have come to know as "male" and "female."

Laqueur, using urine from horses and later human beings, demonstrated that hormones had various roles in the body, even the sex hormones. Testosterone, for example, played a role in regulating women's bodies, just as estrogen and progesterone affected men's bodies in distinctive ways. However, these hormones came to be powerfully associated with gender—in part, Oudshoorn argues, because of the widespread availability of "female" hormones in reproductive medicine.

Oudshoorn charts Laqueur's relationship with the Dutch firm Organon, which was interested in therapeutic uses of "female" hormones, for example in menstrual irregularities and infertility. Testosterone was synthesized in the 1930s, fostering expanded research on the "male" hormone; but without a market for male hormones at the time, efforts focused predominantly on "female" hormones and their application.

Today, "male" and "female" hormones are used in anti-aging interventions for both women and men (Mykytyn 2010), as well as by transgender people to manage bodily functions and appearance. Hormone replacement therapy (HRT) is increasingly marketed as a "fix" for menopause (or the cessation of menstruation), which is framed not as a bodily change but as a disease (Niland and Lyons 2011). Such biomedical interventions are increasingly prevalent among individuals with access to care and economic resources, raising important questions about the role of the pharmaceutical industry (Padamsee 2011).

However, controversy surrounds HRT, with Women's Health Initiative findings suggesting that the risks of combined HRT for aging women outweighed the benefits. In contrast, the Oxford University Million Women Study, funded by Cancer Research UK, was recently deemed to have been "unreliable," unnecessarily alarming women about the effects of HRT, including a reported increased incidence of breast cancer.

Metaphors of regulation

In interesting ways, the body's processes of self-regulation toward homeostasis are also used as resources in classical and contemporary social theory to explain social order and change. As we've discussed elsewhere in the book, the body is a convenient mechanism for thinking, writing, and talking about how humans organize their lives. For example, the father sits at the "head" of the table indicating a paternalistic power position akin to the top of the body.

Sociology, as an academic discipline, is primarily concerned with investigations into how social order is maintained and how social change occurs, with the chief inquiry into who (or what institutions) benefit from the order or change. In the United States, Talcott Parsons (1902–1979) is considered one of the early developers of sociology. Like Emile Durkheim (1858–1917) before him, he believed the individual organism or human body could be used as an instructive model for understanding society; this is called the *organic analogy*.

Each organ in our body is linked to other organs, and they work together to achieve survival through systemic organization. For Parsons, this organic example was similar to social institutions (like the family) being linked to other institutions (like organized religion) to achieve social order through the transmissions and reinforcement of shared norms and values.

Because he was concerned with the maintenance of social order and did not consider or predict massive social changes regarding civil rights, Parsons developed theories primarily concerned with the maintenance of social homeostasis, the crucial feature of all societies (Parsons 1951a). *Social homeostasis* is the system of reinforcing a harmonious social order; if there is a disruption to the functioning of social structures, then social institutions adjust to revert to the familiar.

For example, Parsons theorized the *sick role* (1951b), suggesting that individuals who become ill could temporarily step away from their social duties, engage in help-seeking behavior, secure the appropriate care, and then return to their duties when recovered. In this model, physicians are central to legitimating sickness and also authorizing a release from social duties. Parsons described this arrangement in terms of a "motivational economy" in which the working parts contributed to successful functioning of the whole. The sick role provided a legitimate frame of meaning for what might otherwise be considered "deviant" behavior (e.g., not fulfilling social duties).

According to functionalists like Parsons, who was concerned with the maintenance of social order, the criminal justice system, like health care, is also a social institution that regulates the population through establishing systems of surveillance and "threat" identification. But as we will see in Chapter 14 on Skin, these systems of surveillance are not neutral; rather than creating a social order, they produce and exacerbate existing social inequalities based on race, ethnicity, gender, and social class.

Urban metabolism

The German philosopher, sociologist, and economist Karl Marx used the concept of metabolism to describe material exchanges and interdependent relationships between human societies and cultures on the one hand, and nature on the other. Concerned about unbridled growth of capitalism and the quest for extraction of profit, Marx warned of possibilities of what sociologist John Bellamy Foster (1999) calls "metabolic rift," which

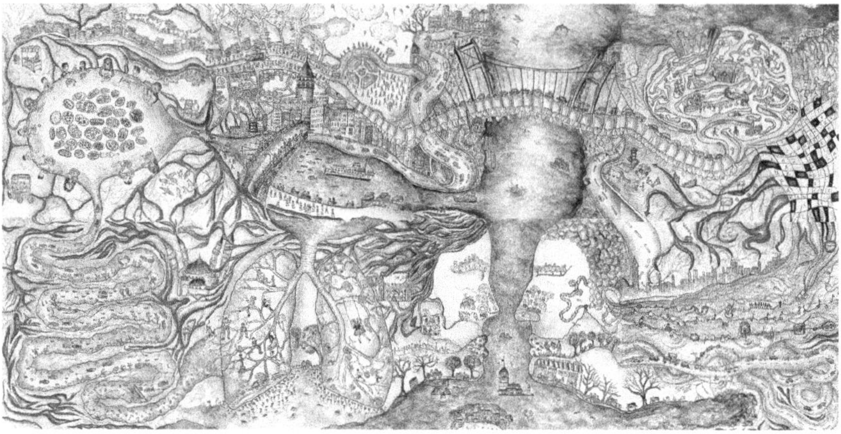

FIGURE 10.1 Istanbul metabolism map
Image courtesy of Deniz Ayaz

describes the social and environmental implications of industrial agriculture and urbanization.

Large-scale capitalist agriculture led to a shift of population from rural to urban areas, creating a "rift" in the metabolism of society. Humans now are said to experience a lack of connection with nature, and this ecological disruption can lead to an unsustainable life, a life without balance, through increased production of waste and ecological degradation. Metabolic rifts in the ecosystem lead to ecological crises, such as climate change or soil loss (Clark and York 2005).

Above is an artist's rendering of the urban metabolism of the Turkish city of Istanbul. Notice the use of anatomical images to depict geographical space and flows of energy.

Urban metabolism uses a bio-physical approach to studying and quantifying urban material and energy flows, drawing on approaches that look to anatomical foundations for industrial ecology (Gandy 2004). The analogies drawn from the metabolic anatomical processes inside our bodies are then applied to the ways goods and energies traffic in and through cities.

The predominant interpretation of urban metabolism is "the study of material and energy flows arising from urban socioeconomic activities and regional and global biogeochemical processes . . . *Urban metabolism* is therefore . . . focused on providing important insights into the behavior of cities for the purpose of advancing effective proposals for a more humane and ecologically responsible future (www.urbanmetabolism.org)."

A CLOSER LOOK

Endocrine disruption

Endocrine-disrupting chemicals (EDCs), including dioxins, phthalates, Bisphenol A (BPA), and organochlorines, may interfere with the endocrine system and production of hormones, which could lead to developmental, neurological, immunological, or reproductive "abnormalities" in humans and wildlife. We absorb EDCs through food, water and air—for example, we are exposed in our everyday interactions and use of household items, our consumption of food and water, and our breathing of dust and particles emitted into the atmosphere.

Synthetic chemicals used in household products, toys, and electronics, as well as commercial and agricultural industries, contain EDCs. These synthetic and man-made chemicals do provide certain quality-of-life benefits to humans in the form of pesticides, insecticides, plasticizers (which make plastics malleable and moldable), antimicrobials, and flame retardants. However, most of these chemicals have not been properly tested to see what their long-term effects are on life forms.

Indeed, aside from chemicals directly added to food, there are *no* mandatory health and safety studies required to put a chemical into commercial use in the United States (Toxic Substances Control Act, 15 USCA 2601–2692). In many regions of the world, there are even fewer regulations (Casper 2003).

The body's exposure to EDCs interferes with its ability to properly distribute and "read" hormones, because essentially EDCs act like hormones. This is particularly dangerous when a fetus or newborn is exposed to EDCs since the hormonal messages trigger the development of tissue and organs in utero (Christensen and Casper 2000). This developmental exposure in early life can lead to significant consequences as a body ages; that is, effects of earlier exposure to EDCs might not show up in an individual until decades after the exposure.

This type of exposure has been termed the fetal basis of adult disease. EDCs have been linked to type 2 diabetes, metabolic syndromes, and obesity (Casals-Casas and Desvergne 2011), as well as to some cancers such as prostate or breast cancers, endometriosis, genital malformations, learning disabilities, and infertility, specifically lowered global sperm counts (Bergman et al. 2012).

A well-known example of such disruption is the case of diethylstilbestrol, or DES. A synthetic estrogen administered to pregnant women from about 1940 until the 1970s, DES was meant to prevent complications. In the U.S., the Food and Drug Administration (FDA) removed DES for use in pregnant women after it was determined that the drug could cause vaginal tumors in girls and women who had been exposed in utero. Males were also found to be affected with such conditions as increased risk of testicular cancer, infertility, and urogenital abnormalities. These "DES daughters" and "DES sons" became the impetus for an *embodied health movement* against DES (Bell 2009).

Population-based studies have been conducted to determine the ways a number of EDCs affect development of reproductive systems and genital organs. One study of Puerto Rican girls used case control methods to examine the differences and potential causes of premature breast development in girls younger than eight years of age, as compared with girls evidencing no breast development. Using blood serum analysis, researchers found an association between plasticizers with known estrogenic and antiandrogenic activity and the cause of premature breast development in a human female population (Colon et al 2000).

Additionally, male genital abnormalities include a condition called hypospadias, in which the opening of the urethra is on the underside of the penis, instead of at the tip. Andrologists Wang and Baskin see a clear linkage between these genital variations and EDCs, "with the increasing incidence of hypospadias paralleling the rate of increase in environmental pollutants, it is imperative that we consider endocrine disruptors as a potential cause for this anomaly" (2008: 503). DES was also found to cause hypospadias (Henderson et al. 1976).

As sociologists Vivian Christensen and Monica J. Casper (2000) demonstrated, the endocrine or hormone disruption thesis activates politics of gender, sexuality, and race, as well as cultural beliefs about the unborn. In contexts in which fetuses, and correspondingly abortion, are highly politicized, such as the United States, the politics of EDCs inevitably intersect with reproductive politics.

Endocrine disorders

There are two types of endocrine disorders. The more common type is a hormone imbalance that occurs when too little or too much hormone is produced. Also, endocrine diseases can occur due to the development of lesions (such as nodules or tumors) in the endocrine system.

Metabolic disorders include acute and chronic illnesses such as diabetes (as discussed in the next section), cystic fibrosis, phenylketonuria (PKU), hyperlipidemia, and gout. Rickets, a common childhood disease in many countries, is a metabolic disorder that results in a softening of the bones and fractures or structural deformities. A team of nutritionists in Bangladesh found that rickets is caused by different environmental and economic conditions.

While vitamin D deficiency causes rickets in areas where either latitude is associated with relatively decreased exposure to sunlight or cultural habits block exposure to sunlight, calcium deficiency has emerged as an important cause of rickets in parts of Africa and Asia, including Bangladesh (Craviari et al 2008).

So depending on where a child lives and their economic stability, rickets could be treated by increasing access to sun exposure or vitamin D in industrialized northern climates or supplementing diets with calcium in poorer regions.

The *heel stick*, lancing a newborn's foot with a narrow tube to collect blood, is most often performed immediately after a baby is born. The blood collected from the heel stick is used to test for congenital diseases. However, in the typical Global North birth story, there seems less discussion of the heel stick than the method of delivery, the baby's first cry, or the APGAR scores (a test delivered to determine the newborn's physical condition). Heel sticks are a taken-for-granted, perhaps invisible medical intervention that possibly registers only when the parents notice the bandage at the bottom of their baby's tiny foot.

Yet sociologist and health advocate Rachel Grob's (2011) study of newborn screening programs in the U.S. reveals the monumental social, cultural, medical, and biological significance of the heel stick. This practice, a state-mandated public health screening, is the most widely used form of genetic testing in the U.S. Grob explains that, since the mid-1960s, when the phenylketonuria (PKU) test was developed, there has been a tremendous increase in testing of babies' blood samples from one disease to over 80 genetic disorders—and counting. Newborn screening worldwide using the heel stick method occurs in 61 countries (Centers for Disease Control 2008).

However, as sociologists Stefan Timmermans and Mara Buchbinder (2012) have shown, while expanded genetic screening in newborns can reveal diseases and save lives, such tests also have a very high false positive rate, potentially frightening new parents. Genetic testing, of both fetuses (Rapp 2000) and newborns, also may offer diagnoses for conditions for which no treatment exists, thus creating ethical dilemmas for parents and healthcare providers.

Diabetes

Diabetes is a chronic condition in which the body does not properly process sugars for use as energy. There are three types of diabetes.

Type 1, also called juvenile onset diabetes, is an autoimmune reaction in which the body's defense system attacks the insulin-producing cells in the pancreas. Treatment of Type 1 diabetes consists of insulin shots to control glucose levels in the body. Only 5 percent of individuals with diabetes have Type 1 diabetes.

Type 2 diabetes is the most common form of the disease. Also called adult onset diabetes, the incidence rate is increasing globally. Type 1 and Type 2 diabetes are both incurable and lead to systemic complications including kidney disease, neuropathy (nerve damage), skin conditions, vision loss, and atherosclerosis (American Diabetes Association 2013).

In fact, diabetes often leads to permanent mobility and visual disabilities, especially in resource-poor settings where treatment is unavailable or limited.

Gestational diabetes, most often an acute rather than chronic condition, is diagnosed in pregnant women after the first trimester when the woman's body is unable to make and use enough insulin needed for pregnancy. Generally, this form of diabetes is treatable and disappears after birth.

Global diabetes prevalence and incidence statistics are startling. Indeed, diabetes is referred to as an epidemic, which is a term typically reserved for infectious conditions. An epidemic means a disease or illness affecting or tending to affect a disproportionately large number of individuals within a population, community, or region at the same time. The greatest number of people with diabetes are between 40 to 59 years of age, which are prime ages for employment. Typically, people with diabetes are facing several decades of managing an expensive and debilitating condition.

Currently, worldwide some 183 million people, or 50 percent, with diabetes are undiagnosed and unaware of their condition. In addition to disabilities, diabetes caused 4.6 million deaths in 2011 of individuals who would have likely lived longer—which epidemiologists call *excess mortality*, or deaths above what would be predicted for any given population. Excess mortality is a significant concern to the stability of a local or regional economy since it may create a "missing generation" of workers and caretakers.

Furthermore, the world prevalence of diabetes among adults (aged 20–79 years) was 6.4 percent in 2010, and will increase to 7.7 percent, or 439 million adults by 2030. Between 2010 and 2030, there will be a 69 percent increase in numbers of adults with diabetes in the Global South and a 20 percent increase in Global North countries (Shaw, Sicree, Zimmet 2010). India, ironically a country often associated with malnutrition, has the highest number of diabetes patients in the world due to its high population rate, genetic predisposition, increased urbanization, and growth in affluence (Ramachandran 2005).

Social understandings of diabetes

Much diabetes research has been driven by the medical model of disease, or an individualistic approach to understanding human behavior and illness. These approaches are limited in that they do not broadly situate the disease in larger social structures or ideologies. In a world of genetically modified or highly processed foods with additives and preservatives alongside the growth of a sedentary labor force, human bodies are consuming and moving differently than in previous generations.

Critical studies of health disparities see diabetes as a prime example of how the medical model of disease is limited. Understanding how systems of stratification deeply influence the prevalence of chronic conditions as well as resource allocation across the globe offers more robust explanations of patterns of disease.

For example, as discussed earlier, *obesity stigma*, or weight-based discrimination and prejudice, can lead to care-seeking barriers for people with diabetes because stigma may cause them to feel ashamed or responsible for their disease (Shafer and Ferraro 2011; Teixeira and Budd 2010).

Sociologists Stefan Timmermans and Steven Haas (2008) explain that:

> Diabetes would appear to be a prototypical case in which to illuminate the social milieu connecting the difficulties of individuals (disease pathology) to larger social problems (the rise in obesity, social disparities in health). Clinical investigations of diabetes are fundamentally limited because they do not address the socially constructed patterns of consumption, lifestyle, and behaviours in which its manifestation in individuals and its great variation between social groups is embedded. Similarly, the traditional sociological study of diabetes as an experiential phenomenon provides little insight into material processes by which the social world comes to be embodied. Disparities in diabetes can only be understood through a sociology that includes the study of disease and its proximate biopathways and that integrates the study of human well being as simultaneously both a biological and social phenomena.

Diabetes and race

There are multiple claims that racial minorities have higher rates of diabetes than whites (e.g. McBean 2004). For example, according to the American Diabetes Association, after adjusting for population age differences, U.S.

national survey data from 2007-2009 for people diagnosed with diabetes aged 20 years or older includes the following prevalence by race/ethnicity:

- 7.1 percent of non-Hispanic whites
- 8.4 percent of Asian Americans
- 12.6 percent of non-Hispanic blacks
- 11.8 percent of Hispanics

Researchers have found that socioeconomic status (SES) is strongly associated with diabetes, perhaps more so or in addition to race and ethnicity (Link and McKinlay 2009). Measuring SES is a challenge for epidemiological researchers because social class is a combination of educational level, income, and occupational prestige. In the case of diabetes, educational level is associated with diabetes rates, meaning that the less education a person has, the more likely they will be diagnosed with diabetes (Smith 2007).

There are several explanations as to why SES affects the rate of diabetes, including access to health care options, affordable nutritional options, ability to engage in health seeking behaviors, such as exercise and relaxation, and knowledge of disease prevention.

In Canada, as in the United States, indigenous populations are especially vulnerable to diabetes. The Public Health Agency of Canada reports that prevalence rates among First Nations, Inuit, and Métis have increased, and that "diabetes is one of the fastest growing diseases among the Aboriginal populations in Canada." The Agency notes genetic, biological, environmental, and lifestyle risk factors, and, although structural, environmental factors are discussed, interventions target management of lifestyle risks such as unhealthy eating and obesity.

Conclusion

While the endocrine system is contained within the human body, in many ways, like respiration, it is a network that links us directly to the environment—all that we eat, breathe, drink, feel, and encounter. In the case of endocrine-disrupting chemicals, our lifelong relationship with the environment is embedded in our flesh, perhaps even in utero, potentially altering the function of hormonal signals across the life course.

Widespread population changes such as industrialization, urbanization, loss of traditional ways, increasingly processed foods, more synthetic chemicals in the environment, a more sedentary lifestyle in certain regions of the world—all of these influence our metabolism. While we may adapt to some changes, others are affecting the very texture of our bodies and lives, leading to increased rates of metabolic and endocrine disease, such as diabetes.

See also

Cells and tissues; Fluids; Bones and muscles; Digestion; Reproduction; Breasts

Key terms

Metabolism, homeostasis, catabolism, anabolism, sociology, organic analogy, social homeostasis, sick role, metabolic rift, urban metabolism, endocrine disrupters, embodied health movements, diabetes, epidemic, obesity stigma, excess mortality

Discussion questions

1. How does metabolism work as a metaphor to explain bureaucracies in everyday life? For example, how are things kept in balance at your university or in your family? What are the types of messages or signals (hormones) that are delivered throughout the campus/family (body)? What is the role of these messages? What does homeostasis look like at your university or in your family?
2. Think about how hormones have been represented and/or discussed in your own life, for example in relation to puberty, aging, and/or bodily enhancement such as athletics training. Are these representations of hormones gendered? If so, in what ways and with what consequences?
3. Considering the longitudinal effects of endocrine disruptors on human bodies, it is likely that your own body has been "interrupted" in normal function by your environmental exposures. The latency period means you might not even be aware of previous exposures and how they are responsible for adult onset illnesses. Discuss the sense of your body as vulnerable. What are some of the ways you take for granted a certain safety in your consumption practices? How is it possible to be "health conscious" in this situation?
4. Historically, epidemiologists called diabetes a disease of affluence. What do you take this to mean? What are some reasons to reconsider this characterization of the disease?
5. If you were designing a study of global patterns of diabetes, what type of questions might you ask to get a broad picture of the trends of the disease? What types of interventions might you explore for disease prevention?

chapter 11

Digestion

Introduction

Morphologically, the human body is a kind of flesh donut. We have a hole—a long tunnel, really—running through our bodies, starting with our mouths and leading all the way down the esophagus, through the stomach and the intestines, and out of the anus. Of course, in anatomical terms, we do not talk about this pathway as the donut hole. Rather, it is the gastrointestinal (or GI) tract, and its proper functioning is key to our survival.

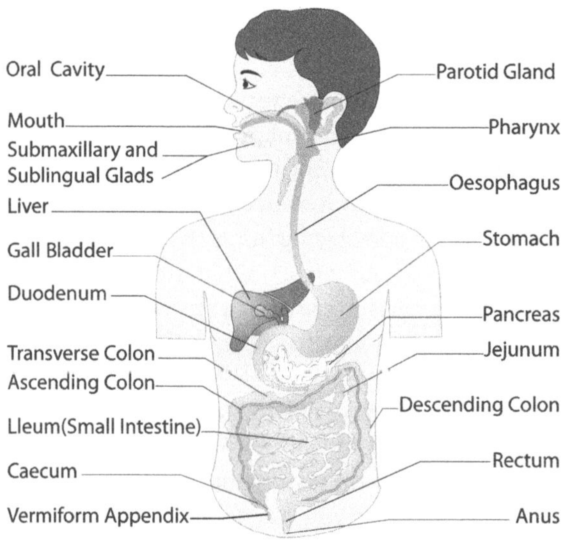

FIGURE 11.1 Digestive diagram
© Shutterstock

The GI tract is divided into different segments: oral cavity, pharynx, esophagus, stomach, small intestine, large intestine, rectum, and anal canal. Accessory glands that contribute to digestion include the liver, pancreas, and salivary glands. The mucosa, or the lining of the mouth, stomach, and small intestine, contains tiny glands, which produce fluids that help to digest food. There is also a layer of smooth muscle along the digestive tract to move food through the system through a wave-like movement called peristalsis.

A tract can mean an indefinitely large area or a major passage of the body. In fact, the GI tract is quite large, with a surface area equivalent to two tennis courts (Jonsson 2013), and when stretched out the digestive tract is 30 feet long! Much of the digestive system is teeming with bacteria and microflora that aid in breaking down food. This massive surface area coupled with our ingestion of foreign matter (e.g., bad bacteria) means that the digestive system plays a significant role in our immune system.

The gut is of growing interest to scientists and the public, with the microbiome a hot topic of biological research. The human body is increasingly framed as an ecosystem for microbes (Pollan 2013), many of which are understood as necessary to human biological processes, such as digestion and development and maintenance of the immune system. Recent research shows that molecules produced by the gut microbiota can affect brain function (Tillisch and others 2013). While fascinating, such findings have been interpreted in reductionist ways to suggest causal links between gut microbes and intelligence, a complex trait.

At the same time, markets have opened up for people to have their microbes tested, raising concerns about privacy and commodification (Stein 2013). For example, for a fee, a person can join the American Gut Project, submit stool samples and a great deal of personal information, and "find out what's in your gut" (americangut.org). Clearly, the gut is compelling, from what we eat to what we expel, including microbes found along the way.

Food, taste, and sociology

Many sociologists study food, food politics, and the work of feeding others and ourselves (e.g., Devault 1994), but this was not always the case. Food and eating were largely viewed as "domestic" or private concerns, and thus not of interest to scholars seeking to explain macro processes, such as industrialization. Yet what and how we eat, and how we make cultural sense of these practices, are deeply sociological (Carolan 2012). Consider that dining together in many cultures is a crucial element of family and community life (Duneier 1994). Or that restaurants in many societies reflect social hierarchies of who cooks, who serves, and who pays to "eat out."

Rural sociologists study social life in rural areas, and as such, have investigated issues related to farming and agriculture, urban-rural-suburban relationships, natural resources, and various dimensions of rural life including socioeconomic and cultural concerns. Sociologists of agriculture, along with scholars in environmental studies, have attended to the politics of food production, distribution, and consumption and to agricultural labor, including race, ethnicity, and migration (Barndt 2008). Food, these studies show, is more than what we eat and encompasses whole systems.

Food also connects to taste and its role in formation of class hierarchies. French sociologist Pierre Bourdieu developed a theory of class distinction based on aesthetics and taste (1984). He argued that taste is a kind of social and cultural capital, and that its formation is a social process that happens over the life course and is transmitted from generation to generation.

Each class has its own criteria for determining what is culturally valuable, or tasteful. What we eat and other everyday "choices" we make are a reflection of culturally specific tastes, with the upper classes eating certain foods in certain ways and the so-called lower classes eating other food in other ways. Food, in Bourdieu's framing, is both a matter of taste (in many senses of the term) and a matter of culture.

A person's tastes, this seemingly individualistic preference, are actually part of that person's *habitus*. Our social position in the stratified world is often illustrated in our *habitus*—what we eat, how we eat, when we eat, and with whom we eat.

Global food politics

Clearly, one of the most significant subsistence needs of any living organism is to feed itself, and human communities unable to feed themselves cannot survive. From recognition of this universal need, the growing interdisciplinary field of *food studies* has emerged.

Food studies examines bio-social systems of food production and use as well as cultural representations. Drawing from the humanities, social sciences, agricultural and nutrition sciences, and public health, the field explores global connections and local alternatives of all things digestible. From a political economy perspective, scholars trace commodities and agricultural concerns from production through consumption, emphasizing international, national, and local food systems. The field also looks at environmental and ethical factors in food production and distribution.

Although consuming food is universal, what we eat and how we eat is not; eating is culturally shaped (Lupton 1996, Kluger et al. 2004). For example, gender can dictate the amount and type of food that a person eats. There are different entitlements to food based on gender, in terms of both quantity and specific food types (Caplan 1997). Women overwhelmingly eat less and are more likely to be vegetarians or vegans. Throughout history, women and children have been denied food as a way of limiting their power (Lupton 1996). Women have also limited their own intake of food to conform to the ideal feminine body type: small, thin, toned, and yet not overwhelmingly muscular (Caplan 1997).

Human ingenuity and communal action have long provided the foundation for solutions to nutritional needs. Food collection, or the reliance on technology and know-how to self-supply food from naturally occurring plants and animals, is not something that is commonly practiced in most "developed" cultures today. However, understanding food collection and the subsistence technologies used to secure food through foraging and hunting is a major contribution of cultural anthropology. Food collection is the oldest human method of finding food and inspired the development of tools, including modifications of sticks and rocks, in order to obtain food more efficiently.

For survival, adults need approximately 2000–2500 calories every day. Men, who are typically larger, are considered to need more calories than women, although this varies widely by region, activity, lifestyle, and so on. Children need fewer calories and also have different nutritional needs as their bodies are continually growing. Human bodies can adapt to a variety of food contexts, but we must have food in order to survive. A body can survive between four to six weeks without food, but only about a week without water. These are material realities of the human body's fragility, and potent reminders that we are both animal and mortal.

World hunger has been a social justice concern since the 19th century. The United Nations Food and Agriculture Organization reports that in 2011–2012, 870 million people globally—or one in eight people—suffered from chronic undernourishment, with children as the most visible victims. Over the past few decades, there has been a paradigm shift in food studies to examining the role that *food insecurity* plays in the everyday lives (and deaths) of the world's poor. Food insecurity means that there is limited or uncertain access to nutritious and safe food; reduced or inconsistent access to food; or irregular food intake. Food insecurity leads to undernourishment, which can weaken immunity, interrupt normal physical and psychological development, and create social disintegration.

Ironically, despite the grave danger hunger presents to human life, there are more overweight bodies in the world than there are malnourished bodies. The World Health Organization reports that, in 2008, there were 1.5 billion overweight people, and the "obesity epidemic," with its own set of politicized health issues, has become headline news in much of the developed world (Guthman 2011). Comparatively, there were 104 million underweight children globally in 2008.

Thus, worldwide some people get too little food while others grapple with food insecurity and genetically modified food.

A CLOSER LOOK

Repurposing feeding tubes

Some modern brides in the Global North, particularly in the U.S. and Europe, have taken to having feeding tubes inserted into their bodies, at a cost of approximately US$1500, in order to lose weight before their weddings.

A study by Cornell University in 2007 discovered that up to 70 percent of brides-to-be want to lose an average of twenty pounds before their wedding day. Techniques for doing so include exercise, controlling food intake, and even surgical treatment, such as tummy tucks and liposuction. However, the extreme body project of weight loss has become increasingly popular among certain brides-to-be. The Feeding Tube Diet, also called the KE or ketogenic enteral diet developed in Europe, promises a 20 lb. weight loss in 10 days.

The ubiquity of the virginal, youthful, demure, pure, and slender (white) bride is an imaginary body that haunts generations of presumably heterosexual (and increasingly, perhaps, lesbian and bisexual) women. What does it mean that in some contexts, girls and women wish not only to be brides, but to be *thin* brides? How might there be different cultural reactions to men pursuing the feeding tube diet in advance of their "big day"? How are these technologies gendered?

The Global Hunger Index continues to remind us of the chronic, widespread problem of hunger in the Global South and many parts of the Global North. What does it say about contemporary global inequalities when some people, such as brides in the U.S. and Europe, can pay to starve themselves through innovative uses of a feeding tube, while other people, particularly poor and vulnerable children in poorer nations, are being starved by structural violence, unequal distribution of resources, and domestic and foreign policies that advantage the wealthy?

Processed and convenient

The ways in which humans have processed fresh food—whether recently slaughtered animals or produce just plucked from the soil—have shifted across time, affecting bodies. For much of history, humans preserved fresh food by drying, fermenting, or freezing it. The National Center for Home Food Preservation states that, as early as 12,000 BC, cultures in the Middle East and Asia dried food by leaving it in the sun. In climates with frigid temperatures during at least part of the year, people could store food outside. The first "refrigerators" were holes packed with snow and ice. Now, in the developed world almost every home is equipped with a refrigerator that also includes a freezer.

Industrial canning has come to almost entirely replace home food preservation and has made possible a longer "shelf life" for fruits, vegetables, and some meats (e.g., seafood). It has also made possible long-distance shipping. Whereas a Mason jar full of home-canned plums might shatter en route if shipped, aluminum-encased peaches or sardines can be boxed into cartons and shipped almost everywhere by plane, train, or truck, with little damage. The same can of food, once purchased by a consumer, can sit on a pantry shelf for months, if not years. The advent of industrial canning has relegated home canning to an "artisanal" activity in many parts of the world.

Alongside canning also came industrial freezing, with all sorts of food being wrapped (sometimes pre-cooked) into packaging for mass distribution. Consumers can purchase fruits, vegetables, meat, desserts, and indeed entire meals right out of grocery or supermarket freezers, take them home to store in their own freezers, and then pop them into the microwave or oven when they are ready to eat. In the early to mid-20th century, frozen "TV dinners" changed how families—and especially women—prepared meals and also how families consumed them. Some families migrated from the dining room table to the living room sofa as televisions also entered households.

Canned and frozen goods have long been marketed as convenient, presumably for those preparing the food (most often women). Indeed, canned and frozen foods are among the only types, along with "junk food" (e.g., chips, nuts, cookies, sodas), sold by modern-day convenience stores that also sell items such as cigarettes and magazines. In the developed world, and especially in the United States, whole neighborhoods are defined as food deserts. These are areas, usually impoverished, lacking in fresh fruit, vegetables, and other healthful foods. Food deserts are characterized by a lack of grocery or farmer's markets, and instead tend to have access only to convenience stores.

A great deal of effort has gone into convincing consumers that canned and frozen goods are just as good as fresh. But as geographer Susanne Freidberg (2010: 17) argues: "We've come to see freshness as a quality that exists independent of all the history, technology, and human handling that deliver it to our plates—a quality that, ironically, transcends time and space precisely because it is so sensitive to both." That is, an extraordinary amount of work has been expended by a great many industries and people to make food that is not actually fresh seem as if it is. Here, processing is equated with both nutritional value and convenience.

Reflecting one of the immense contradictions of the modern world, in many developed countries fresh food such as butchered meat and organic local produce is consumed by middle- to upper-class people, while canned and frozen goods are largely consumed by the poorer classes. In the Global South, where hunger is a chronic problem, there is sometimes fresh food, such as food farmed or butchered in communities, but far less often processed food. All too often, there is no food at all.

Slow food

The stratified nature of food production and consumption has led to a "return" to artisanal farming and food production in the developed world, or what has come to be known as the slow food movement.

Founded in Italy in 1986, the slow food movement has become institutionalized in the organization Slow Food International, whose logo is a snail. With chapters, or "convivia," in over 150 countries, the organization's manifesto states that slow food is an escape from "the tediousness of 'fast food'" and "fast life," and the cultivation of taste and "a better quality lifestyle." Slow food is "an idea, a way of living, and a way of eating."

Described as a grassroots movement, the slow food movement advocates for greater biodiversity, helps to foster a network of small-scale farmers and food producers, engages in food and "taste" education, and connects consumers and producers. Emphasis is on locally grown, fresh, healthy food; for example, slow food advocates participate in campaigns against genetically modified organisms (GMOs) and for seed preservation. The key to slow food is sustainability, of both food resources and human and animal communities.

Emergence of the slow food movement in the Global North must be situated historically and politically, in the context of the industrialization of food production including processed foods, factory farming, genetic modification, chemical agriculture, food insecurity, and other practices viewed as unsustainable and damaging to the environment, animals, and human health.

Factory or corporate farming is a particular target of slow food advocates, who instead wish to encourage local, small-scale, artisanal farming. In the U.S., for example, less than 1 percent of the population are farmers. In 2012, the total number of farms in the country fell to 2.17 million, a record low. The industrialization of agriculture raises numerous consequences for critics: excessive pesticide use, lack of price controls, abusive working conditions for immigrant laborers, threats to food security and food sovereignty, collapse of rural communities, empowerment of industrial lobbyists to shape national policy, and damage to family livelihood.

Factory farming also invokes ethical concerns about the treatment of animals produced as food. Massive cattle, pig, and chicken operations have been investigated by journalists and government officials who report horrific living conditions, cruel methods of containment and slaughter, and often unsanitary environments (Kirby 2011). Animal advocate and author Temple Grandin (2010), who is also autistic, has worked to improve the living and dying conditions of animals produced for food, turning her insights from autism to the greater good.

"The new domesticity"

The movement for slow food, artisanal life, and recovery of the forgotten home arts such as pickling and knitting has sparked considerable debate about women's roles in the Global North. Publication of Emily Matchar's *Homeward Bound: Why Women Are Embracing the New Domesticity* (2013) generated lively dialogue on various social media sites and raised the hackles of some feminists. After a century of activism to enable women to venture outside of the home and into workplaces and public life, why would they want to go back?

Matchar writes: "Clearly, something has been shifting in our culture over the past decade or so. The various pieces—the urban chickens, the domestic-porn blogs, the retro cookery, the attachment parenting—are beginning to come together to reveal a larger whole . . . [O]ur current collective nostalgia and domesticity-mania speak to deep cultural longings and a profound shift in the way Americans view life. I call this phenomenon 'new domesticity'" (2013: 4). Other commentators have termed these newly homebound women "feminist housewives."

But what does it mean, for women and men, to be "newly" domestic in the developed world, to have bodies that stay at home to raise chickens and nurture infants? Especially relative to rural and non-developed settings where subsistence living (e.g., raising animals, growing food, sewing one's own clothes) is the norm. The "new domesticity" taps into some dissatisfaction with "modern" living, including industrialized food production and the rapid pace of life.

But perhaps, too, this nostalgic longing signals a new direction in Global North feminisms: where can modern, industrialized women go next? And if it is back into the home to reengage the womanly arts, should feminists worry that this is another way of appropriating the lives and practices of women in the Global South?

A CLOSER LOOK

Genetically modified foods

GMOs, or genetically modified organisms, have emerged as one of the most important and controversial topics in transnational food justice movements. A GMO is a food substance, such as rice, corn, or fish, that has been genetically engineered for a certain purpose. For example, edible plants can be made resistant to pests and drought and thus produce a higher yield, or may be modified with the addition of important nutrients, such as rice with vitamin A.

Indian environmental activist Vandana Shiva (2000), among others, has argued against GMOs and globalization, suggesting that food production must be diverse, with local and global practices of seed sharing (and not seed patenting), and that it also must be in the hands of local farmers and producers, and not multinational corporations. Shiva is a major proponent of biodiversity, as for example through her work with Navdanya, a women-centered NGO committed to environmental and cultural sustainability.

In the Global North, a "right to choose" movement has advocated for better labeling of food, such that GMOs are identified at the point of purchase. Several companies, including chemical giant Monsanto, have faced considerable anti-GMO sentiment, threatening their bottom line (e.g., profit). Whole Foods Market Inc., purveyor of organic and fine foods, announced that, by 2018, it would require the labeling in its stores of any genetically engineered food.

However, advocates of GMOs argue that genetic engineering has the potential to feed many more people, and thus to reduce global hunger. Proponents also point out that humans have always manipulated food through agricultural and husbandry practices, and GMOs are simply the next step in a long history of such interventions. These claims may be somewhat specious, however. For example, one pro-GMO website is backed by Monsanto, DuPont, and Dow AgroSciences.

What *is* the future of food, and of the human bodies reliant on food?

Food pathogens

Despite major advances in food storage and preservation, food can be contaminated in multiple ways to make it unhealthy for human consumption, threatening wellbeing and survival.

Bacteria are single-celled living microorganisms ranging in size from one micrometer (one millionth of a meter) to 10 micrometers in length. They are naturally found in the environment (often in a spore form) or in various animal reservoirs. Bacteria can multiply in or on food and cause foodborne infections in persons who consume contaminated food or liquids.

Toxins most often associated with foodborne illnesses are poisons produced or released by certain bacteria, such as salmonella. When ingested, bacterial toxins usually act locally within the human body, but may spread to other parts and damage cells, tissues, and the host immune system, as for example in the paralytic disease botulism.

Viruses are minute organisms that reproduce only within living cells. Nonetheless, they can remain infectious in food and may cause foodborne infections in humans. Hepatitis A virus and Norovirus (formerly known as Norwalk-like virus) are the most recognized food-related viruses.

Parasites are single or multi-celled organisms with dimensions greater than 10 micrometers. Like viruses, some parasites reproduce within host cells, while others can contaminate food and water. However, many parasites develop a cyst form that is inert and resistant to the environment. This cyst, when ingested through food or liquids, can multiply within humans and cause foodborne infections.

Alarmingly, many of the "advances" that have made food plentiful and accessible for many in the Global North (but not all people; hunger rates are still very high) have also contributed to the emergence and distribution of pathogens. In several instances, contaminated food can be tied directly to unsafe food practices or industrial practices with little to no regulatory oversight. When food becomes subject to the assembly line, there are many more points of entry for pathogens into our food and our bodies.

Diseases of the gut

The digestive system, with its many moving parts, is an ideal environment for the eruption of disease. From "tummy aches" to colon cancer to GERD (gastroesophageal reflux disease), often what ails us is either already in our intestinal system or introduced there through eating and drinking or infection. Here, we discuss a few diseases that affect the digestive system.

Cholera is an acute diarrheal infection caused by ingestion of food or water contaminated with the bacterium *Vibrio cholerae*. Annually, there are an estimated three to five million cholera cases globally, which lead to 100,000–120,000 deaths. The bacterium's short incubation period, of two hours to five days, enhances the potentially explosive pattern of outbreaks. Epidemiologists state that cholera is *endemic* in many countries, which means that the disease is constantly present to greater or lesser extent in a particular locality. One would expect to find a consistent rate of the disease, and thus it is managed according to that rate. It is worth noting that cholera can be treated very simply with *clean* drinking water, which would prevent almost all deaths.

Celiac disease is an immune disorder in which the digestive system is damaged by the body's response to gluten. Gluten is a protein most commonly found in wheat, rye, and barley. People with celiac disease have difficulty digesting the nutrients from their food and may experience diarrhea, abdominal pain, bloating, exhaustion, and even depression when they eat foods with gluten. In the past 50 years, the prevalence of celiac disease has quadrupled in the United States as well as increased internationally.

The increase is attributed, in part, to changes in infant feeding from breastfeeding to the introduction of baby formula with increased gluten content, as well as infection (Tack et al. 2010). The foods we feed babies and nursing mothers matter for the subsequent development of allergies and illnesses.

Treatment for celiac disease includes dietary exclusion of all gluten products and supplementing nutrition with iron, calcium, and vitamins. Gluten-free diets are, while helpful, prohibitive for many because "speciality" products can be expensive or difficult to find, and gluten contamination can occur in packaged foods and foods prepared at home.

Clostridium difficile, or *C. difficile*, is a spore-forming bacterium that lives inside of our bodies. It is usually held in check by the "good" bacteria in our systems. However, if the "good" bacteria are reduced in number, for example through use of antibiotics, *C. difficile* may proliferate. While sometimes treatable, it can lead to diarrhea, inflammation of the colon, toxic megacolon, and even death. The Mayo Clinic in the United States reports that: "C. difficile infections have become more frequent, severe and difficult to treat." Infectious disease specialist Torbjorn Noren (2010) reports *a tenfold increase* in *C. difficile* in the Global North.

As with many diseases, structural issues play a major role in the emergence and distribution of disease. Here, we see contaminated food and water, hospital practices, and excessive antibiotic use as factors in illness.

Elimination

What goes in must come out. Food is no exception, with elimination of waste a key component of our digestive systems. Yet when food comes out, it typically looks nothing like it did when it went in, whether it emerges as feces, urine, or vomit. Our bodies process food from start (e.g., chewing and swallowing) to finish, breaking it down into its nutritional components along the way. What the body does not use for energy and other needs, it excretes.

Whereas eating and dining are highly social activities, often taking place in families or public settings, elimination of waste—as discussed earlier—is typically private, at least in geographic areas where private or minimally shared bathrooms are the norm. Indeed, houses in many "developed" countries have two or more bathrooms with running water and toilets. This is not the case in poorer parts of the world, where lack of sanitation facilities and practices is a major public health concern. By one estimate, 2.6 billion people around the world do not have access to toilet facilities.

Chronic diarrhea is a major problem in the Global South and is directly related to lack of adequate sanitation facilities and clean water. The World Health Organization reports that 1.8 million people die annually from diseases related to diarrhea, and that 90 percent of these are children under the age of five. Improved water supplies and sanitation can reduce these deaths significantly.

In the Global North, anorexia and bulimia are considered important public health issues. Anorexia nervosa is a disease characterized by low (sometimes extremely low) body weight, an obsession with thinness, and non-normative or disordered eating habits including self-starvation and induced vomiting. Bulimia nervosa, another form of disordered eating, involves binge eating and purging through vomiting or use of diuretics and laxatives to induce waste.

Anorexia and bulimia, conditions largely of developed nations where food is plentiful and available, predominantly affect young women. Along with other disordered eating, such as extreme dieting, these practices can be understood, in part, as a response to cultural ideals of beauty, body type, and womanhood (Fallon, Katzman, and Wooley 1996). There is some evidence that disordered eating is increasing in countries/regions as they become more developed (Makino, Tsuboi, and Dennerstein 2004), raising interesting questions about the role of the media and hegemonic cultural messages.

A CLOSER LOOK

Bathroom/toilet practices for trans and disabled bodies

Elimination, or the process of excreting human digestive waste, is a major role of the digestive system. This act is universally human and it is often met with social taboos as it is considered dirty,

FIGURE 11.2 Inclusive bathroom sign
Image courtesy of Thomas Bogart, 2014

private, and shameful (Cavanaugh 2010). Furthermore, in many cultures there is strict maintenance and adherence to gendered norms about how to deposit our digestive waste. The creation and regulation of public restrooms, washrooms, and bathrooms have emerged to force dichotomous practices in toileting that separate men and boys from women and girls.

Thus, gender variant people may face rampant and routine discrimination, harassment, and violence when accessing gender-segregated public bathrooms (Toilet Training 2003). A U.S. survey of 6,450 transgender and gender non-conforming people found that 26 percent of respondents in educational settings reported denial of access to facilities including bathrooms, 22 percent of respondents reported being denied access to appropriate bathrooms at work, and 21 percent reported being unable to find or organize a suitable bathroom situation at work (Grant, Mottet, and Tanis 2010).

But changes are afoot in designing and labeling public bathrooms. In 2009, Thailand's Kampang Secondary School created a third restroom for "ladyboys" for the 10 percent of students who identify as transgender (Maclean 2009). Many college campuses now feature gender-neutral bathrooms and unisex bathrooms, as well as traditional men's and women's stalls, as do some restaurants and other public spaces.

Compliant bathrooms, or bathrooms that adhere to standards of accessibility for people with disabilities, are also not easily located, especially in countries with little to no regulation regarding disability access and/or universal design. This lack of accessibility to compliant public restrooms for people with disabilities leads to stressful situations and often makes travel impossible for those in wheelchairs. The United Kingdom and the United States are especially notorious for being inaccessible (e.g., tube stations with no lifts, subways with no elevators), as are many countries throughout the Global South.

Thus, just as food consumption is stratified, so too is food elimination. Some bodies do not "fit" into standard, socially accepted spaces set aside for urination and defecation.

Conclusion

In certain contexts, to digest means to assimilate information and organize in a systematic fashion. People digest massive amounts of information every day through visual, auditory, and social cues. We process these cues, which inform our interactions. From a sociological perspective, somatic digestion begins with food production and follows food as it moves through distribution, procurement, preparation, ingestion, and elimination.

This chapter examines our digestive system as it spans outside of our bodies into the geopolitical and biomedical environments that restrict, constrain, and/or condition our ability to obtain or grow healthy, good, and affordable food. In some cultures, extreme body projects are undertaken by individuals attempting to regulate digestion as part of deliberate impression management strategies.

Simultaneously, in a world of such vast differences in wealth, having access to nutritional food is hampered by neoliberal ethics of profit extraction. Food is plentiful and nutritious for some, while many others starve.

See also

Cells and tissues; Organs; Fluids; Endocrine system and metabolism; Faces

Key terms

Food studies, food insecurity, slow food movement, food desert, endemic

Discussion questions

1 Take a random sample of food advertisements from your everyday life. How are men and women portrayed differently in these advertisements? What are their body sizes? Are they racialized? What type of food are they consuming? How are they animated in their consumption? How does this food consumption and advertising reinforce or resist prevalent ideas about gender?
2 Have you ever preserved food? If so, in what context? If not, why not? What do you think of the "new domesticity"? Does a return to traditional practices offer women more or fewer choices?
3 If antibiotic use has been linked to the development of diseases such as C. *difficile*, how might we address the epidemic? Should we eliminate use of antibiotics? Find newer and better antibiotics? Change hospital practices? What can be done?
4 Global health is deeply stratified, giving us pervasive hunger and diarrhea in the Global South and diseases such as anorexia and bulimia in the Global North. What can we learn by comparatively addressing these? Is disordered eating, such as anorexia nervosa, even a possibility in, for example, sub-Saharan Africa, with high rates of child starvation owing to lack of food?
5 Have you ever been mistaken for using the "wrong bathroom" or have you ever corrected someone to use a different bathroom? What cultural work is being performed in disciplining individuals to use their sex-segregated toilets? Are digestion and elimination gendered practices? If so, what makes them gendered?

chapter 12

Reproduction

Introduction

Every species on the planet, including humans, must reproduce or die out. Like other mammals, the member of our species categorized as "female" gestates offspring in the womb, most often singleton fetuses but sometimes twins, triplets, or other multiples. Unlike most animals but similar to some other mammalian species (e.g., dogs, nonhuman primates, livestock), human beings reproduce through an array of social, technical, and institutional interventions.

These interventions—ranging from assisted reproductive technologies (ARTs) to 4-D ultrasound images to pitocin drips during childbirth to cesarean sections—are a long way from the "natural" reproduction of previous eras (when mortality rates were also higher). While they offer new ways for humans to reproduce (or not, if one uses contraception), they also have immense potential to alter women's and men's reproductive experiences as well as social and cultural meanings of reproduction.

Numerous processes and issues can be located under the broad rubric of "reproduction," including abortion, contraception, adoption, infant mortality, sterilization, pregnancy, childbirth, genetic testing, miscarriage, stillbirth, surrogacy, sperm donation, breast milk banking, and more. This chapter focuses specifically on the *reproductive body*, or the ways in which human bodies—and especially women's bodies—are manipulated in the service of propagating the species.

It is important to note that while both women and men are involved in reproduction in unique ways that rely on various body parts, across many cultures reproduction is understood and represented as primarily an issue for and about girls and women. This quite monolithic representation has consequences for how societies understand and intervene in human reproduction, as well as how reproduction and childrearing are legislated given women's near-universal marginalization and inequality in relation to men.

The mechanics of reproduction; or, what's under the hood?

In bodies typically categorized as female, the reproductive organs include the uterus (also sometimes called the "womb"), the fallopian tubes, and the ovaries. Genital organs including the vagina and cervix are also often included in descriptions of the female reproductive system.

The ovaries produce eggs from egg cells present at birth; girls are born with about 400,000 egg cells in their ovaries. Once generated through follicle stimulation, the mature egg (or, very rarely, eggs) then travels via the fallopian tubes to the uterus; this process is called ovulation. If unfertilized, the egg exits a woman's body during menstruation as the uterus sheds its lining.

If an egg is fertilized in the oviducts or uterus through contact with sperm, a zygote is formed. In healthy or "normal" pregnancies, the zygote implants itself in the wall of the uterus where it develops into an embryo and eventually a fetus.

In an ectopic pregnancy, the zygote implants outside of the uterine cavity, for example in the fallopian tubes, cervix, or abdomen. Such pregnancies are recognized in most settings as medical emergencies. Untreated ectopic pregnancies can lead to bleeding, miscarriage, and—in more serious cases—death of the pregnant woman. Surgery may be required to remove the embryo.

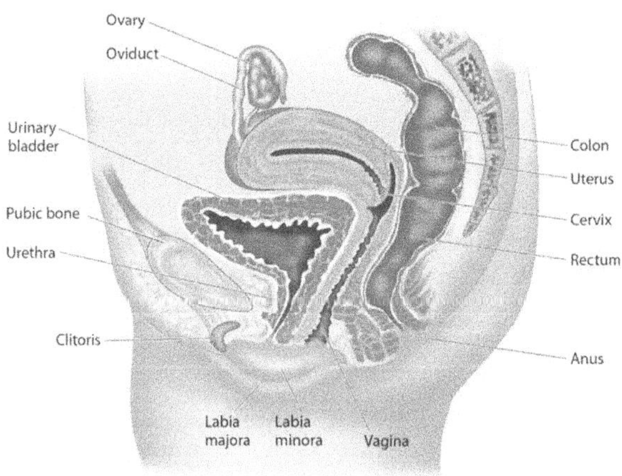

FIGURE 12.1 Female reproductive system
Image courtesy of Sinauer

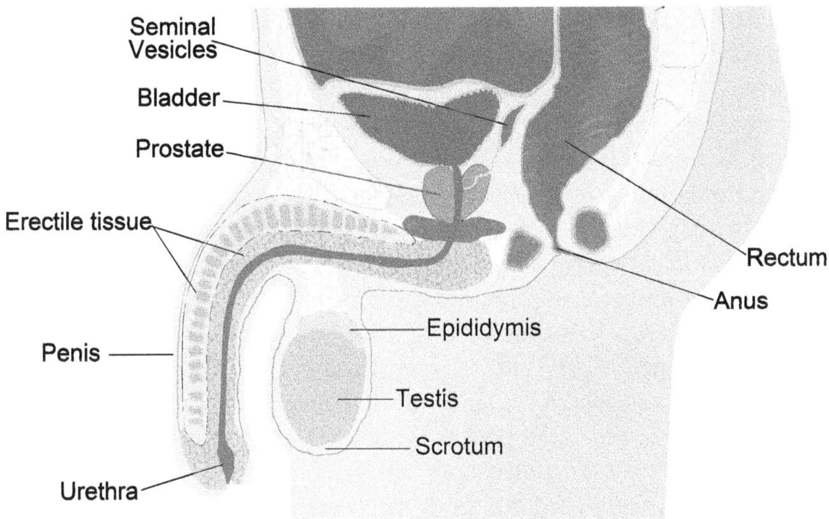

FIGURE 12.2 Male reproductive system
© Shutterstock

In bodies typically categorized as male, reproductive organs include the testes, penis, epididymis, and vas deferens (or sperm duct). The testicles, contained within the scrotum, generate sperm, which take 70 days to mature. During sexual activity, the penis may harden as it is stimulated; during ejaculation, sperm are released in semen that flows from the penis. If semen is deposited into a fertile woman's body, pregnancy may ensue.

Gendering body parts

Like "female" and "male" genital anatomy described in Chapter 17, reproductive anatomy has also been conceptualized in terms of gender differences, including typical gender roles.

For example, anthropologist Emily Martin wrote about how the egg and sperm are frequently narrated in scientific textbooks in terms of a fairy-tale love story, with sperm embodying normative male characteristics and eggs assuming normative female characteristics. She notes also that menstruation is framed in textbooks as a kind of reproductive failure (where pregnancy is the presumed goal of all female reproduction), and male reproduction—including the prolific production of sperm—is described in "breathless prose" (Martin 1991: 486).

Similarly, Lisa Jean (Moore 2007) has written about sperm as central to conceptions of men's reproductive agency, sexual power, and identities as fathers, sexual partners, men, and citizens. The term "sperm" accrues multiple meanings depending on the context of its use; for example, compare the happy, father-like "tadpoles" of children's "facts of life" books to the villain's DNA in sperm samples from rape cases. The most dominant definition of sperm is found in biological descriptions of human reproduction. However, this biological definition is completely enmeshed with socio-cultural meanings.

Prenatal exposure to tobacco has been shown to lower sperm counts in male offspring. Excessive alcohol consumption, marijuana smoking, and obesity also affect sperm counts and sperm performance. Endurance bicycling has been shown significantly to alter sperm morphology, or shape, while masculine performances of risk-taking such as military action or dangerous work can have detrimental effects on vital sperm production (Moore 2007). Researchers have called for studies to determine if storing cellular phones near the testicles in the front pocket of pants might decrease semen quality.

Owing to the sexual, cultural, and personal significance of ejaculation, men and boys are vulnerable to prevalent messages about the timing, size, and frequency of their ejaculation. Ejaculation can be a visual representation of the measure of a man (Moore and Weissbein 2010). In addition, ejaculation is most commonly preceded by an erection, a biological process rampantly medicalized. For example, premature ejaculation is the labeling of rapid ejaculation whereby the penis ejaculates sooner than desired. Allopathic treatment as well as over-the-counter interventions proliferate, including impulse-buy supplements at the local drug store and homeopathic remedies at health food co-ops.

These gendered representations of reproduction, while reflective of many Western perspectives, are not universal. Anthropologists Robbie Davis-Floyd and Eugenia Georges (1996) report that in the Basque region of France, conception is understood as similar to cheese making; and among the Hua people of New Guinea, conception is believed to result from combining semen with menstrual blood. Davis-Floyd and Georges report that in the three dominant monotheistic religions—Islam, Judaism, and Christianity, all of which are male dominated—the male role in reproduction is considered creative (e.g., sperm as seed production) while the female is represented as the passive "fertile soil" into which seeds are sown.

In short, although we understand a great deal about how reproduction works, biological knowledge and the "facts of life" are deeply shaped by cultural meanings. What we know about "the egg and the sperm" reflects how societies are organized—typically unequally—around gender roles and perceived differences between men and women.

Stratified reproduction

Anthropologist Shellee Colen, writing about West Indian childcare workers in New York City, coined the term *stratified reproduction* in 1986. With this concept, she argued "that physical and social reproductive tasks are accomplished differentially according to inequalities that are based on hierarchies of class, race, ethnicity, gender, place in a global economy and migration status and that are structured by social, political, and economic forces" (1986: 78). Her ethnography of childcare workers evidenced the ways that caring for children by immigrant women reflected social inequalities between the West Indian caregivers and the women (and men) who employed them.

For more than two decades, Colen's term has had significant longevity. It has been used to analyze issues from assisted reproductive technology (Davis 2009) to medical services for infertility (Greil et al. 2011) to the outsourcing of surrogacy (Maranto 2011). Indeed, reproductive experiences of all kinds are shaped by a woman's social location.

In the U.S., for example, this has meant a radically different history of reproduction for African American women than it has for white women, with coercion, violence, and diminished autonomy a hallmark of Black women's experiences (Roberts 1997). In many other countries, the experiences of women marginalized by race, ethnicity, class, migration status, age, and other factors has led to more coercive interventions, such as forced sterilization in China (Greenhalgh 2008) and the coercive use of Norplant in Egypt (Morsy 1986).

As sociologists, we find the term "stratified reproduction" immensely helpful for thinking about a range of reproductive issues and also about reproductive bodies. Which bodies are allowed to reproduce? Which bodies—human and nonhuman (e.g., livestock, zoos)—are forced to reproduce? Which bodies are prevented from reproducing? Which bodies bear the physical marks of coercive medical and governmental intervention? Which bodies matter in policy formation?

These questions are fundamentally about power, social inequalities, and *governmentality* as applied to human bodies. From theorist Michel Foucault, we understand governmentality to mean the techniques and rationalities through which subjects are governed. Reproductive processes, as Foucault demonstrated (1976 [1990]), are a prime locus of governmental practices.

Reproductive processes

Reproduction, like most bodily processes, is inherently social. It takes at least two, and often many more, people to see a baby into the world or to prevent conception in the first place. The key individuals involved in reproduction, and particularly the women whose reproductive processes are managed from start to finish, engage in reproductive processes corporeally *and* socially, and often politically.

In addition to the individual woman or couple, there are numerous other actors behind the scenes also working to mediate human reproduction in a multitude of ways: reproductive scientists (e.g., Clarke 1998), obstetricians (e.g., Drife 2002), midwives (e.g., Gaskin 2011), geneticists (e.g., Franklin and Roberts 2006), test subjects (e.g., Briggs 2002), abortion providers (e.g., Joffe 1996), fertility specialists (e.g., Mamo 2007), sonographers (e.g., Taylor 2008), women's health advocates (e.g., Davis 2007), a variety of healers the world over (e.g., Jordan and Davis-Floyd 1992), fetal surgeons (e.g., Casper 1998), state agents (e.g., Browner and Sargent 2011), sperm donors (e.g., Moore 2007), egg donors (e.g., Ameling 2011), gestational hosts (e.g., Markens 2007; Twine 2011), and many more.

In what follows, we explore a range of reproductive processes—contraception, sterilization, miscarriage, abortion, fertility, pregnancy, and childbirth—in terms of how they engage the body, who is involved, and sociological considerations.

Contraception

The term "contraception" literally means against or counter to conception. Also called birth control or family planning, contraception refers to methods and devices used to prevent pregnancy during heterosexual intercourse, or that between a man and a woman.

Contraception, far from being a modern development, was used in antiquity (Bujalkova 2007). In ancient Greece, pessaries made of linen,

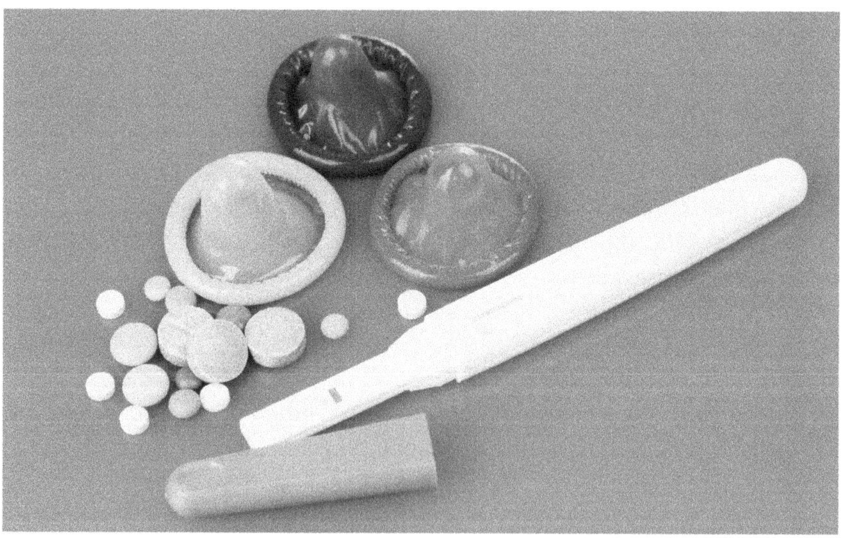

FIGURE 12.3 Contraception
© Shutterstock

sponge, or wool and prepared with wine, honey, rose oil, goose fat, resin, and other substances were used as contraceptives and *abortifacients* (abortion-inducing). In ancient Rome, women lubricated the mouth of the uterus with honey, resin, white lead, or myrtle oil, and often inserted wool into the uterus to prevent conception.

According to PBS: "'Womb veils,' a 19th-century phrase for diaphragms, cervical caps, and condoms, often made from linen or fish intestines, have been in use for centuries. In the 1700s, the famous seducer Giacomo Casanova told of using half a lemon rind as a cervical cap." Historian Andrea Tone (2002) traces use of a variety of technologies in early America, including pessaries and douches, prior to development of the more widely used oral contraceptive.

Coitus interruptus, also known as "withdrawal," "pulling out," and "natural family planning," was mentioned as early as the story of Onan in the Torah. In this "ancient" method still in use today among heterosexual couples, a man removes his penis from a woman's vagina prior to ejaculation. The method is useful for people with little access to or an aversion (for example, religious beliefs) to contraceptive technology, but it is much less effective than many other methods.

Contraceptives that block sperm from encountering the egg are called barrier methods; they include male and female condoms (Collier 2007), cervical caps, diaphragms, sponges, and chemical spermicides (agents that kill sperm). Barrier methods avert conception by preventing the merger of egg and sperm, thus interrupting the "fairy tale romance" described by Emily Martin (1991). Some barrier methods (e.g., condoms, dental dams, microbicides) also help to prevent transmission of sexually transmitted infections.

Perhaps the most revolutionary form of contraception, especially for women, was the Pill (Watkins 2001), widely viewed as ushering in a new era of sexual freedom. Described as "a marvel of modern chemistry," oral contraceptives work by preventing ovulation; a woman ingests hormones contained in the pill, which interrupts the body's "normal" functioning. When a woman wants to become pregnant, she stops using the contraceptives.

The birth control pill exemplifies what sociologist Adele Clarke (1995) describes as a historic shift in bodily interventions from modern control to postmodern transformation. Unlike barrier methods, which block the merger of egg and sperm through technologies such as latex, hormonal methods, including the Pill, DES, Depo Provera, and Norplant, fundamentally alter the body's chemistry in the service of preventing conception. Bodies are transformed from the inside out; women's bodies primarily, as the development of male contraception, including a "male pill," has been stalled in large part for social and cultural reasons (Oudshoorn 2003).

Since 1982, the most common methods of birth control in the United States have been the Pill and sterilization, and globally the most common are permanent (female sterilization) or long lasting (the IUD). However, economic factors are at play here, too, as pharmaceutical companies seek additional, transnational markets for their products. Sociologists Laura Mamo and Jennifer Ruth Fosket (2009) analyze the ways that Seasonale, an extended-cycle oral contraceptive, has been marketed for uses other than contraception, including menstrual suppression and skin care.

Why would a woman (or a man) want to prevent pregnancy? Many reasons, including the ability to enjoy sexual activity without the threat of becoming pregnant or impregnating somebody. Contraceptives, to some degree, "equalize" men and women, by enabling women to engage in sexual intercourse without reproductive repercussions. Of course, this very ability has fostered significant changes in women's status and that autonomy is often threatening to men, especially in more conservative societies (Saleem and Bobak 2005).

Contraception also allows couples the world over—where they have access to technologies—to plan how many children they will have, the spacing of their children, and to avoid having children at all. In societies where family planning is widely used, women have higher status, attend school longer, and participate in the labor force at higher rates. Health benefits include lower rates of maternal morbidity and mortality and a reduced number of abortions (Ahmed et al. 2012).

And yet, contraception is globally stratified. Women in the developed world have access to a range of technologies that have become safer and more effective than ever before, making birth control statistically safer than pregnancy (Hess 2013). However, even in wealthier nations such as the United States and Canada, women of color and First Nations women face starkly different reproductive "choices" and futures than do white women (Silliman et al. 2004; Roberts 1998).

In poorer countries, many women still have very limited options, including exposure to those technologies most conducive to coercion such as Depo-Provera (McDonough 2013). Given these limitations and consequently higher rates of unplanned pregnancies, rates of maternal morbidity and mortality are significantly higher in the Global South than they are in most of the industrialized world, as evidenced by the World Health Organization.

Sterilization

Sterilization is a permanent form of contraception and is the most widely used method globally to prevent pregnancy. The procedure does not involve introducing any hormones into the body but does involve biomedical (including surgical) intervention. It is most often used by people who do not want additional (or any) children—although it is sometimes promoted and used in coercive ways (Clarke 1984).

In women's bodies, sterilization involves closing or blocking the Fallopian tubes through various methods. The most invasive form, surgical tubal ligation, involves cutting or suturing the tubes. Considered to be major abdominal surgery, in this procedure patients must be anesthetized.

In recent years, less invasive non-surgical procedures have become available. For example, the Essure method involves placing nickel coil inserts into the Fallopian tubes; over a three-month period (during which contraceptives must be used to prevent pregnancy), scar tissue forms around the inserts, creating a permanent blockage.

Touted as a major breakthrough when it first appeared in the U.S., Europe, and Canada, Essure has now become mired in controversy as women have come forward to report problems and to seek reversal of their sterilizations.

This is interesting relative to another procedure, quinacrine sterilization, which has been quite controversial since its inception. A method that uses acid to burn and scar the tubes, quinacrine has been used predominantly in women in the Global South, often coercively by governments and with the support of Global North organizations and providers. Quinacrine illustrates some of the worst consequences of colonialism applied to reproductive bodies (Flanagan and others 2013).

In men's bodies, sterilization is typically referred to as a vasectomy. Considered minor surgery that can be done in a doctor's office or clinic, it is much cheaper than female sterilization. In vasectomy, the vas deferens is severed, thus preventing sperm from entering into the ejaculate. Vasectomy is considered permanent, but it may be reversible.

Although cheaper and less invasive than tubal ligation, fewer men undergo vasectomy than women undergo tubal ligation, illustrating the extent to which preventing pregnancy is still viewed as a woman's problem.

Sterilization is popular worldwide, but it also has been subject to considerable abuse in practice. *Compulsory sterilization*, also called involuntary sterilization, has been used in many countries, usually in conjunction with *eugenics* programs, or those attempting to encourage reproduction by desirable groups while discouraging reproduction of "undesirables."

For example, Hitler incorporated sterilization practices into his Aryan nation-building endeavors, targeting women and men suffering from alcoholism, Huntington's disease, epilepsy, and mental illnesses (Proctor 1988). Involuntary sterilization has been a key strategy in enforcing China's one-child policy (Greenhalgh 2008).

In the United States, women of color were sterilized frequently without their knowledge or consent (Davis 1981), as were the mentally ill, prisoners, and other marginalized populations (Kluchin 2011). Disabled girls and women have been frequent targets of compulsory sterilization, and many HIV+ women have been sterilized without their consent in South Africa, Namibia, and Kenya (Mojtehedzadeh 2013).

An especially egregious example of compulsory sterilization is described in the film *La Operación*. Beginning in the 1920s, a coalition of U.S. government officials, Puerto Rican officials, and medical professionals began to push for mass sterilization in Puerto Rico, viewing overpopulation as a cause of social problems. Demographer Harriet Presser (1969) reported that a third of women had been sterilized, many in private birth control clinics; this marked Puerto Rico as having the highest percentage of sterilized women anywhere in the world. In a stunning display of embodied colonialism, Puerto Rican women were later used as research subjects for oral contraceptives (Briggs 2002).

Thus, sterilization—while providing many people worldwide with a permanent form of contraception in line with their own reproductive choices—has also been used to *deny* reproductive autonomy for the most disadvantaged members of societies—often with the full support of the state.

Miscarriage and abortion

Miscarriage and abortion both involve the end of an existing pregnancy, but with some very crucial differences that contribute to how these reproductive processes are viewed socially and legislated by the state.

Miscarriage is the spontaneous or unplanned loss of an embryo or fetus before it is capable of living independently, at or before about 20 weeks' gestation. There can be many causes of miscarriage, and sometimes women may not even know they have miscarried if it happens early enough in gestation. But often women *do* know, and with a planned pregnancy, a miscarriage can bring tremendous grief and a sense of loss (Layne 2002). (With an unplanned pregnancy, it may bring relief.)

While the prevalence of miscarriage indicates that it is a common occurrence—more than one in five pregnancies ends in miscarriage—it is fairly uncommon for miscarriage to be broadly acknowledged or discussed (Casper and Moore 2009).

The Mayo Clinic reports the following risk factors for miscarriage: maternal and paternal age, previous miscarriages, chronic conditions, uterine or cervical problems, substance use (e.g., alcohol, drugs), and invasive prenatal tests. Other factors that can negatively impact pregnancy include domestic abuse (Alio et al. 2009), environmental hazards (Brown 2008), malnutrition (Mahmood et al. 2011), chromosomal anomalies of the embryo/fetus (Simpson 2007), and various forms of structural violence (Peters 2012).

An *abortion* is the selective or deliberate termination of an embryo or fetus. Laws governing abortion are a patchwork and vary by country, with some nations banning the procedure entirely, others allowing it only in cases where the health of the pregnant woman is threatened or the pregnancy is caused by rape or incest, and still other nations allowing it "on demand" within the first trimester.

Primary technologies include surgical abortion, such as manual vacuum aspiration, and medical or medication abortion, which uses a combination of drugs (i.e., mifepristone, methotrexate, and misoprostol). With all abortion techniques, the earlier in pregnancy the procedure is performed, the safer it is.

Globally, despite abortion's varying legal status, women who experience unplanned pregnancies and who do not wish to remain pregnant attempt to secure abortions, and many die doing so (Hadad and Nour 2009). For this reason, among others, women's health advocates around the world have pushed for safe, legal, and accessible abortions.

Pregnancy and birth

Most human pregnancies last about 40 weeks, although some babies may be premature (i.e., delivered early) or late, up to 42 weeks and beyond. (By contrast, hamsters carry their young for 15-17 days, while elephants carry theirs for almost two full years.) In general, pregnancies lasting more than 42 weeks are considered "high risk" or non-normative, and thus induction and/or cesarean section may be performed where these services are available and accessible.

It is a hallmark of the women's health movement to recognize that some women receive too much medical care, while others too little. Pregnancy and birth exemplify this pattern.

Women in many Global North settings typically experience pregnancy as highly medicalized, with tests, diagnoses, fetal monitoring, drugs, and other interventions. The U.S., with a for-profit health care system, exemplifies the high-tech approach, where delivering babies is big business (Margulis 2013). In such settings, attempts to undergo "natural" childbirth with non-physician providers such as midwives are often viewed as alternative, radical, and even criminal.

Cesarean sections, while potentially life-saving for some women, have increasingly been framed as an "epidemic" with researchers worldwide attempting to address what many view as excessive rates of intervention (Betrán et al. 2007). In the U.S., about one in three women gives birth via cesarean section (Hamilton et al. 2011), in China about half of women give birth via cesarean section (Hvistendahl 2012), and in Brazil, a whopping 80–90 percent of women in private hospitals have c-sections. C-sections contribute to elevated rates of maternal morbidity and mortality (Liu et al. 2007), as well as to adverse outcomes for infants (MacDorman et al. 2006).

However, at the same time that high-tech pregnancies and births abound, some women in the Global North—and many women in the Global South—receive little to no prenatal care and little to no intervention, even when it is urgently needed. Poor women in many countries labor and deliver their babies without the gleaming "bells and whistles" available in well-stocked hospitals in the United States, Canada, Europe, and parts of Latin America. Many women the world over undergo labor and give birth in very low-tech (or no-tech) settings accompanied only by midwives, lay practitioners, family members, or traditional healers.

It is ironic that some women in the Global North desire less intervention and have sought home births, delivery in birthing centers, midwifery care, doulas, and other "natural" services. While at the same time, birthing women in the Global South are dying at astonishingly high rates—up to 800 women per day, 99 percent of them in poorer nations—for lack of critical health care services and due to structural factors including poverty, inequality, and lack of empowerment. Advancing women's health through reduction of maternal mortality is one of the United Nations Millennium Development Goals—and one that has been especially resistant to success.

Preterm birth—that is, pregnancies that end before a fetus is fully developed—contributes to adverse maternal and neonatal outcomes, and rates are highest in Global South countries, especially Africa (Lawn and others 2010). Preterm labor and birth are more prevalent among younger mothers (e.g., those in their teens whose bodies are not adequately developed to sustain a baby), malnourished women and girls, and those with other health complications.

Consequences of preterm birth may include mortality or complications in preterm infants, neurodevelopmental issues including motor impairment, and cognitive disabilities (Institute of Medicine 2007).

A CLOSER LOOK

Obstetric fistula

Obstetric fistula, an injury prevalent in the Global South and rare in most of the Global North, is typically caused by unattended obstructed labor. Fistulas are tears in the body's organs. Severe tissue damage and fetal death can result from prolonged labor, but this is largely preventable in the Global North, where adequate health care is more readily available. In the poorer nations, there are up to 100,000 new cases of obstetric fistula annually.

Among the world's poor, rural women who have little access to specialized health care, obstructed or prolonged labor can cause a fissure to develop between the vagina and bladder (known as vesicovaginal fistula) or between the vagina and rectum (known as rectovaginal fistula), or both. After a case of prolonged labor, the fetus or baby usually dies. The woman, if she survives, is left chronically incontinent.

The physically disabling features of fistula may include a crushed pelvis, damaged internal organs, and chronic leakage of urine and/or feces. Affected women suffer from bodily odors deemed offensive, and are considered to be unclean. Because the genitals are part of the injury, family members may believe women to have a sexually transmitted disease. Women are prevented from fulfilling their culturally-defined roles as wives and mothers, and are often socially ostracized. Women whose babies die at birth are also grief-stricken at the loss.

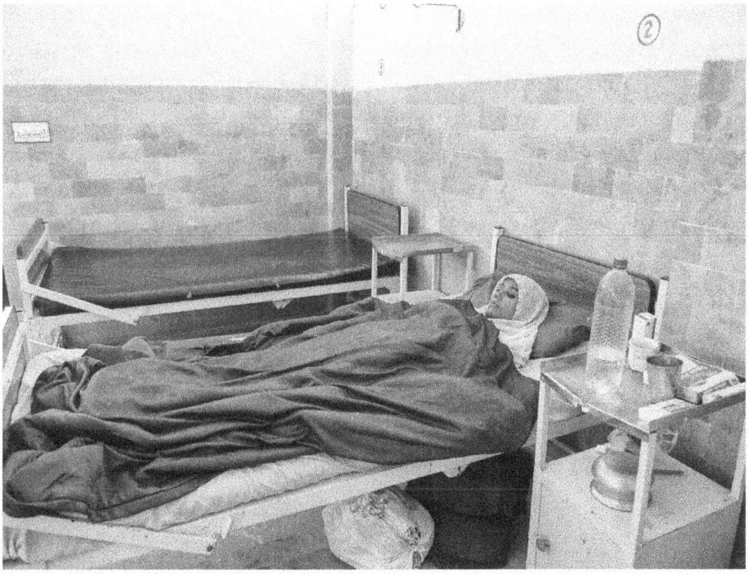

FIGURE 12.4 Female incontinence
© AFP/Getty Images Photo taken by Asif Hassan/Stringer: Razia, a fistula patient resting at her bed at Koohi Goth hospital in the outskirts of Karachi.

What makes obstetric fistula especially tragic is that it is *both preventable and treatable* where health care services are available. Such injuries reflect the lack of vital services for the world's most vulnerable people:

"poor, young, illiterate girls and women in remote regions" (de Bernis 2007). Girls in cultures where they are married young are especially vulnerable, as their bodies are not adequately developed (e.g., their hips are not wide enough and their bones not strong enough) to sustain pregnancy and childbirth. As Ahmed and Holtz (2007: S14) state: "having a fistula changes a woman's quality of life forever."

Transnational women's health NGO EngenderHealth reports that up to 90 percent of obstetric fistulas can be repaired surgically. A growing movement of NGOs, medical providers, donors, and local governments is targeting fistula, partly because improved maternal morbidity and mortality rates are part of the United Nations Millennium Development Goals. Clinics have been established in regions of the Global South, and trained practitioners are traveling there to help with repairs. Alongside health services, social support networks are being established to help counsel women and their families, and to reintegrate women into their communities.

However, obstetric fistula must also be *prevented*, which will require structural changes and economic investments in obstetrical care in the Global South, family planning services (including provision of contraception *and* abortion), and policy changes including eradication of child marriage. The United Nations Population Fund has initiated the Campaign to End Fistula, emphasizing prevention (including eradicating poverty), treatment, and education to help destigmatize the condition.

Reproductive politics

Of all the human bodily processes, reproduction is among the most politicized. This is because, as we have shown, sexuality and gender are deeply implicated in reproduction. That is, what societies believe about the comparative roles of men and women—and their social worth—is projected onto the ways human beings reproduce.

Women's bodies very often become the target of state interventions and governance strategies, such as *pronatal policies* (encouraging reproduction) and *antinatal policies* (discouraging reproduction). Governments also may collude with healthcare professionals in the "best interests" of embryos and fetuses and/or of the state, thereby limiting pregnant women's reproductive autonomy (Browner and Sargent 2011; Paltrow and Flavin 2013).

In different ways but with similarly adverse effects, the reproductive policies of China and Romania offer examples of the social costs of intervening in women's reproductive lives.

China's one-child policy was instituted in 1979 to limit population growth while the government implemented various economic reforms. The policy relies on universal access to long-term contraceptives, sterilization procedures, and abortion (Hesketh et al. 2005). Families that violate the policy face heavy fines, and women who become pregnant may face forced abortions. Exceptions include rural families in certain areas and families whose first child is disabled or a girl—indicating a clear preference for nondisabled and male children.

Widely regarded as repressive and even "brutal" (Jian 2013), the policy has had several major consequences. First, China faces a sex-ratio gap with more boys than girls; considerable global attention has been paid to the problem of China's "missing girls" (Saletan 2009). The ratio of old-age dependency has also increased, such that China is facing an increasing number of elderly and an increased ratio between elderly parents and adult children (Hesketh et al. 2005).

With forced sterilizations and abortions as well as allegations of female infanticide, the policy has come under fire as a violation of women's human rights. Jian (2013) reports that while wealthy families are able to circumvent the policy, the country's poorest women are most vulnerable to abuse and coercion. She writes:

> Almost every one of the pregnant women I spoke to had suffered a mandatory abortion. One woman told me how, when she was eight months pregnant with an illegal second child and was unable to pay the 20,000 yuan fine (about US$3200), family planning officers dragged her to the local clinic, bound her to a surgical table and injected a lethal drug into her abdomen. . .It is not surprising that China has the highest rate of female suicide in the world. The one-child policy has reduced women to numbers, objects, a means of production; it has denied them control of their bodies and the basic human right to determine freely and responsibly the number and spacing of their children.

China's policy exemplifies antinatalism, or the active discouragement of reproduction for at least some segment of the population. In contrast, in 20th century Romania, Nicolae Ceausescu's regime exemplified a pronatalist stance.

Concerned about declining population rates, in 1967 Ceausescu criminalized abortion with few exceptions while also working to position motherhood as women's highest calling (Kligman 1998). Decree 770, as it was known, required gynecological exams and other intrusions into women's lives, and many delivered babies they were ultimately unable to care for. Others, desperate, sought to terminate their pregnancies; between 1965 and 1989, more than 9,000 women died from illegal abortions (Kligman 1995).

After the Ceausescu regime fell, thousands of children were abandoned to hospitals and orphanages where they were subjected to horrifying standards of living (Cohn 2011).

Conclusion

Reproduction is intimately linked to human survival; without replicating ourselves biologically, the human species would die. Although reproductive processes are material and corporeal, they are also deeply and necessarily social. Reproduction, often framed as an individual "choice" or behavior, is rather one of the most social of human bodily processes, bringing a variety of social actors together.

As we have shown for a number of reproductive processes, human reproduction is also political and contested. Women's bodies, in particular, are targeted for state schemes and often intervened in by government and biomedical actors in ways that diminish women's autonomy. Pregnant women are especially vulnerable to coercion and harm.

These politics reflect longstanding, persistent hierarchies of gender, race, and class, suggesting that the reproductive body—almost always framed as a woman's body—is also a socially marginalized body.

See also

Genes; Endocrine system and metabolism; Breasts; Genitals

Key Terms

Reproductive body, stratified reproduction, governmentality, abortifacient, compulsory sterilization, eugenics, miscarriage, abortion, pro-natal policies, anti-natal policies

Discussion questions

1 This chapter shows how thoroughly human reproduction is gendered. Can you talk about some reasons why this might be the case? Why is reproduction, more than say metabolism or digestion, politicized? How does this play out in your country or region?

2 Women use contraception far more often than men, there has been minimal progress in developing male contraception, and men are less likely than women to elect sterilization. How do these patterns reflect social dynamics? How is male privilege enacted through contraceptive use?
3 While abortion and miscarriage both involve the end of a pregnancy, there are crucial differences. Discuss these differences, especially with respect to women's status. Should there be more public discussion of these reproductive processes? Why or why not?
4 Governments in nations such as India welcomed quinacrine sterilization, as it promised to offer a cheap and accessible option for managing population growth. However, after women's health advocates exposed problems with quinacrine, it was banned for use. Discuss the relationship between a nation's need for population control programs and women's reproductive autonomy.
5 How is obstetric fistula both a medical problem and a social problem? What are some possible solutions to obstetric fistula?

PART III

What others see

chapter 13

Faces

Introduction

Faces are a highly visible and expressive body part, perhaps the most obvious marker of human recognition. Located prominently on the front of the head, the face is quickly noticed when encountering another human being. We talk of "looking like" other people, whether genetically related or not, and when we do so, we are usually referring to the face.

Most babies visually learn their mothers' faces quite rapidly. Pre-verbal children take in the social cues of their parents and others through facial expression, and as the first "smile" attests, we celebrate proper facial socialization with positive reinforcement. (Visually disabled babies learn their mothers in other ways, for example through smell and touch.)

Faces lend themselves extremely well to metaphor: facing forward, saving face, losing face, a long face, facing off, having face. These metaphors suggest a profound connection between faces and social identity, a connection that numerous scholars have explored. Home to the eyes, ears, nose, mouth, and complex muscles and nerves, the face is the corporeal "seat" of identity and expressed emotion, and thus of great interest to sociologists and others including poets and artists, who have long been interested in representing human faces through prose, images, and three-dimensional forms such as sculpture.

Science writer Carl Zimmer (2008) reports that naturalist Charles Darwin was fascinated by faces. In 1872, Darwin produced a volume called *The Expression of the Emotions in Man and Animals*, in which he suggested that expressive faces were a product of evolution. He argued that facial mimicry and higher-level expression of emotions evolved as humans did, with the mammalian face emerging as an important tool for communication.

For many animals—human and nonhuman—faces are a major key to social recognition. Humans, chimps (Parr 2011), elephants (Plotnik, de Waal, and Reiss 2006), and dolphins (Reiss and Marino 2001) have the ability to recognize themselves in the mirror, and thus to recognize "kin" of their species. Even crows have been found to recognize individual human faces (Nijhuis 2008). It is presumed, however, that humans, far more than other species, tend to rely primarily on visual recognition of self and others.

Faces are essential in nonverbal communication, which operates largely on the basis of visual cues such as facial expression, eye contact, posture, and gesture (Andersen 2007). Without spoken language, the body becomes an instrument for communicating, as in sign language. However, what works in one culture to convey meaning nonverbally may not work in another (Levine and Adelman 1992). Context matters in how facial expressions and other gestures are perceived. For example, although smiles are considered universal, in some settings people do not typically smile at strangers.

Given the centrality of faces to social life, those that are damaged, nonnormative, injured, or just look "different" can be disruptive to social order; they can impede social recognition and acceptance (Talley 2014). Faces can become the contested fleshy terrain on which battles for inclusion are fought. Facial interventions, whether cosmetic surgery or burn repair or skin whitening or more mundane technologies such as wrinkle cream, are often about overcoming marginalization and facilitating social inclusion.

In this chapter, we explore sociological ideas about faces, beginning with Charles Horton Cooley's "looking glass self" and then moving on to Erving Goffman's theory of "face-work." We next take a closer look at sociologist Heather Laine Talley's (2014) work on facial surgical interventions, including face transplantation.

Then we turn to the eyes, ears, nose, and mouth, showing multiple ways in which these body parts are socialized. We offer a second closer look, this one focused on Botox, and then close with a discussion of the role of facial recognition in 21st century biosecurity practices and facial reconstruction in forensic anthropology.

The looking-glass self

Charles Horton Cooley was an American economist and sociologist who lived from the 1860s through the first third of the 20th century. He was the author of numerous articles and books, including *Human Nature and the Social Order* (1902). His most well known concept is "*the looking-glass self.*"

Drawing on the heuristic of a mirror, or looking-glass, he theorized that a person's self-image is created through interaction with others, and in particular through our sense of how another perceives us. Cooley's theory includes three major components: (1) imagining how our self appears to others; (2) imagining what others think of our appearance; and (3) developing feelings and/or a sense of self on the basis of others' perceived judgments.

A classic example of Cooley's theory is a person practicing a "speech" or "delivery of information" in front of a mirror as they try-out different ranges of facial expression. As human beings, we want to manage social reception through control and presentation of our facial nuances.

Cooley was interested in social organization, social processes, and subjectivity, and the concept of "looking-glass self" enabled him to explain interactions between people as well as self-identity. In Cooley's formulation, the basis of psychological identity was social, created through interaction with others. His work offered an important structural corrective to psychological theories that located the self solely within the individual mind or deriving from biology. Decidedly sociological, Cooley offered a theory that considered the face as a socially managed and culturally situated tableau for the delivery of interactive messages.

Face-work

Erving Goffman, a Canadian sociologist, theorized "face-work" as central to social interaction. Like Cooley, Goffman was interested in social organization and is known for a number of important sociological concepts, including *stigma* and *total institutions*. Much of his work was aimed at explaining how human beings relate to one another within social systems.

In *Interaction Ritual: Essays on Face-to-Face Behavior*, first published in 1967, Goffman defined "face" as "the positive social value a person effectively claims for himself by the line others assume he has taken during a particular contact" (p. 5). For Goffman, a "line" is defined as "a pattern of verbal and nonverbal acts," a kind of social action. The corporeal face, where feelings are expressed visually, is the resource used in the production and maintenance of social face. Feeling pride and entitlement can then be expressed through our faces and then reflected back to us as a sense of positive accomplishment of the reception of messages.

During interactions, people behave according to the sense of "face" they believe they have on the basis of others' reactions, much like Cooley's looking-glass self. However, Goffman theorizes beyond the construction of such "face" to the ongoing maintenance of social face. People act confidently when their face "matches" the social code or order. And yet, Goffman suggests, people can also be "out of face" or "in the wrong face" if their expressive behavior does not conform to the code. For example, consider the consequences of somebody smiling or giggling during a somber occasion, such as a funeral or a religious ceremony.

This formulation connects self-identity to social organization, in that once people offer a positive self-image to others, they are under social pressure to maintain that self-image, and to prevent themselves through various rituals from being "out of face." Subjective feelings of insecurity, shame, embarrassment, and confusion must be managed by, ideally, projecting a face of confidence and leadership.

Being "out of face" can also refer to disfigured or non-normative faces, as we discuss next.

A CLOSER LOOK

Face transplantation

In her pioneering book *Saving Face: Disfigurement and the Politics of Appearance*, Heather Laine Talley (2014) offers the first sociological analysis of face transplantation. Drawing on ethnographic research, she provides a sophisticated empirical grounding for a new sociology of the face.

In 2005, surgeons in France completed a partial face transplant, setting off a storm of media coverage. Several institutions began clinical research programs exploring face transplantation—which raised a host of ethical issues. For example, face transplantation patients must commit to lifetime use of *immunosuppressant drugs*, bringing questions of risks and benefits to the fore.

Comparing face transplantation with three other sites of facial intervention (male-to-female transgender feminization surgery, Extreme Makeover, and Operation Smile, an international charitable organization that offers free cleft surgeries), Talley explores the ways in which surgical repair is used to address *disfigurement*. However, she also calls into question the social conditions under which a face is deemed "disfigured," arguing that to be disfigured is as much a social as a corporeal condition. In other words, it is our collective reaction to any particular face that is deemed "disfigured" rather than a permanent definition of what a disfigured face is.

Talley shows the ways in which disfigurement matters in people's lives. Those deemed to have faces that are disfigured, ugly, or otherwise non-normative encounter consequences ranging from social exclusion to on-the-job discrimination to significant challenges to interpersonal relationships. Many of the subjects in her study note that disfigurement can be "worse than death." Hence, the imperative to "fix" the face technologically.

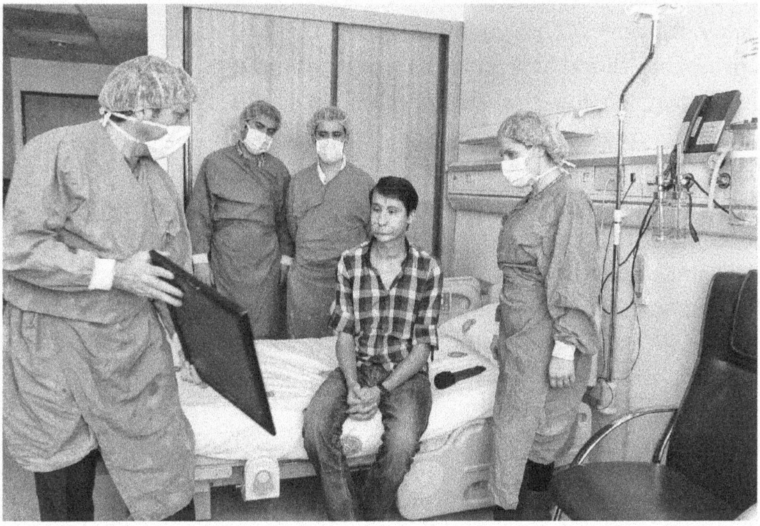

FIGURE 13.1 Face transplant
© *Photo by Akdeniz University-Pool/Anadolu Agency/Getty Images)*
Turkey's 7th face transplant surgery is performed at Akdeniz University Hospital in Turkey's southern province of Antalya on February 15, 2014. During the operation, which took four hours and ten minutes, the face of Ali Emre Kucuk (not seen) is successfully transplanted to 22 year-old Recep Kaya (C), who has already gone under about 40 operations since he blew up a part of his face in an accident while he was playing with his father's hunting rifle nine years ago.

With respect to face transplantation, a practice in which most often a damaged face is "repaired" with tissue from the patient's own body (which produces better outcomes than a skin graft from a deceased donor), the intervention creates a dichotomy between "vital" and "nonvital" repairs. Increasingly, many problems with appearance are deemed severe disfigurement, regardless of the scale of damage, and thus eligible for "life-saving" repair.

Talley situates her analysis of face transplantation firmly within the interactionist tradition of Goffman and the interdisciplinary field of disability studies, offering a 21st century example of "face-work" in multiple, high-tech, and increasingly transnational settings. While three of her research sites are predominantly in the Global North, Operation Smile is international, and indeed reflects considerable Global North economic and moral investments in the world's children.

Eyes and visuality

Sociologist Everett C. Hughes (1971) described "the sociological eye" as a way of looking at and analyzing the world around us, likening it to C. Wright Mills's (1959) sociological imagination. In Hughes's framing, the sociological eye was not limited just to academic sociologists; anyone could develop an analytical lens on the world. Thus, the concept is democratizing rather than exclusionary. Anyone can learn to use their sociological eyes to interpret the world around them and discern patterns in social relationships, such as inequality.

As Hughes's concept shows, visuality and the eyes pervade sociology, for example through methodological practices such as participant *observation* and *ethnography*. Indeed, the discipline was founded on the idea that one could *see* and thus interpret social life. However, until recently, sociology had little engagement with the interdisciplinary field of visual studies, despite the example of visual anthropology as a distinct field (Banks and Ruby 2011). Only recently has *visual sociology* emerged as an area of study in its own right. Visual sociology investigates visual aspects of society and social life, for example advertisements, photographs, art, and material culture.

Despite this lacuna in sociology, what we understand about the eyes and visuality is relentlessly social and political, from how we see (Berger 1972), to who and what we see (Casper and Moore 2009), to the spatial politics of visual impairment (Ginley 2013), to plastic surgery among Asians seeking more normatively "Western" eyes (Zane 2001), to the proliferation of visual images (Croteau, Hoynes, and Milan 2012). Theorizing the visual is an important and inherently sociological project.

Consider, for example, eye contact. While in some cultures people are socialized to make frequent eye contact as a measure of sincerity and purpose, in other cultures, eye contact is to be avoided. Whereas eye contact might mean friendliness in one setting, in another it could be read as a challenge to altercation. Further, who can make eye contact with whom is stratified by gender, age, race, ethnicity, status, and other social factors. In some cultures, women who initiate eye contact may be perceived as inappropriately sexually promiscuous.

Eyes are also the sites for certain diseases and conditions, including cataracts (clouding of the lens), glaucoma (damage to the optic nerve), retinal disorders (e.g., diabetic retinopathy), macular degeneration (thinning or bleeding of the macula), and conjunctivitis (or "pink eye"). Many of these are age-related; that is, as we grow older our eyes weaken and become less functional, affecting our vision.

Eye conditions and diseases are very common, and also largely treatable in settings where people have access to care. Yet the World Health

Organization reports that 51 percent of the world's blindness is caused by cataracts, affecting some 20 million people with limited or no access to health care. Trachoma, another prevalent disease, is caused by the organism *Chlamydia trachomatis* and spread through contact with the person infected. WHO estimates that trachoma affects 21.4 million people, causing visual impairment in 2.2 million people and blindness in 1.2 million people.

Living in a highly visual world means that people with impaired or no vision may face numerous obstacles, although they may also have rich, full, interesting lives. Many people around the world—WHO estimates 285 million—have some form of visual impairment requiring correction through glasses or contact lenses. Approximately 39 million people are blind, and another 246 million have low vision. WHO reports that 80 percent of vision problems are preventable and correctable—with access to care.

In the United States, blindness is considered a disability under the Americans with Disabilities Act and thus warrants legal protection and accommodation in education and the workplace. The U.N. Convention on the Rights of Persons with Disabilities affirms non-discrimination, equality of opportunity, and accessibility for people with disabilities, including blindness. And yet, the International Labour Organization finds that in some countries, unemployment of people with disabilities, including visual disabilities, is as high as 80 percent.

Ears and aurality

Ears, when they are working properly, are conduits of sound to our brains. As part of the body's auditory system, they also help us to balance and locate ourselves in space. Anatomists divide the ear into three parts: the outer ear, the middle ear, and the inner ear. The relation between these shapes how we hear sound, or not, and how we move through the world.

Ears are more than appendages for hearing; they are also a primary location for bodily adornment. People have pierced, decorated, and reshaped their ears for centuries for religious, ethnic, political, sexual, and other reasons including marking important life passages. In some cultures, ear adornment is so common that to have a bare ear is considered deviant; however, in others cultures a single ear piercing might be considered a radical statement about gender or sexual orientation.

The link between ears and sound extends beyond the physiological. We speak of musically gifted people as having "an ear" for music. We speak of actors as having "an ear" for dialect; American actress Meryl Streep is famous for her ability to hear and repeat accents. Sometimes, we refer to people unable to grasp cultural differences as "tone deaf," referencing the physiological/musical inability to hear relative pitch.

Glaswegian Steve Goodman (2009), an expert in music culture and a DJ, has written about sonic warfare. In his research, he explores how sound has been used as an acoustic weapon to produce discomfort and pain among targets. For example, the U.S. Army aimed high-decibel sound at Manuel Noriega and the U.S. Federal Bureau of Investigation targeted the Branch Davidians in Waco, Texas. In his framing, sound becomes a tool for crowd control. Just as piped in music can be used to soothe and calm crowds at shopping malls or dentist offices, loud, repetitive, or atonal sound can be used to agitate and attack.

Controversy surrounds the U.S. Navy's use of sonar, which disrupts the hearing of marine mammals such as whales and dolphins. Because these animals rely on sound for communication, to orient themselves, and to secure food, human-induced noise is more than just harmful; it can be lethal. Here, we see "sonic warfare" waged in the world's oceans, with devastating consequences for the bodies and lives of marine mammals.

Diseases of the human ear range from ear infections (most common in babies and children, but also affecting swimmers and others), to tinnitus (a roaring in the ears), to Menière's disease (resulting from fluid in the ear). These conditions can affect hearing as well as balance, with symptoms such as dizziness and lightheadedness and even temporary or permanent hearing loss. Historically, deafness was associated with "dumbness" or mental deficiency, and the deaf were ostracized, institutionalized, or worse. Deafness was a target of 19th and 20th century eugenics campaigns including those in Nazi Germany (Ryan and Schuchman 2002).

According to the World Health Organization, 360 million people suffer hearing loss. With regular check-ups and intervention, about half of all hearing loss could be prevented. (Hearing loss is also age related.) However, many people around the world do not have access to basic hearing screening or technologies for repair and assistance. WHO estimates that only about 10 percent of people who need hearing aids currently have them.

Deaf culture is organized around the needs, meanings, practices, lives, and practical concerns of people who are deaf and who use sign language. The term "Deaf culture" is typically a positive one, denoting community, pride, and inclusion. However, Deaf people and communities have had to respond to decades of stigma, misunderstandings, and exclusion from mainstream normative culture (Bauman and Murray 2013).

Today, the use of cochlear implants to "fix" deafness has become mired in contention, with some Deaf advocates eschewing the technology and others choosing it (Mauldin 2012). The Academy Award-nominated documentary film *Sound and Fury* (Aronson 2000) chronicles this complex issue in one extended family.

Noses and olfaction

As a child, did you ever pinch your nostrils together to avoid a "bad" taste, such as a vegetable you didn't like or cough syrup? This technique works, in part, because noses are part of our digestive system, affecting taste and nutrition. Noses are also part of the olfactory system, which is directed toward sensory perception of odors, and the respiratory system, which keeps us alive through respiration. Through cilia and mucus in the nasal passages, noses protect our bodies from pathogens, dust, and other potentially harmful intruders.

Thus, with our noses, we breathe, taste, smell, and manage the world around us.

As we all know, not everything smells (and tastes) delicious or appealing. Having noses means that we are also exposed to less savory odors, such as rotting garbage, animal carcasses, spoiled food, and possibly toxic emissions, depending on where we live and work. Often, "bad" smells can alert us that something is wrong, for example the smell of smoke leading to discovery of a fire.

The abundance of smells that surround us (and that we produce) means that socially, we spend quite a lot of time managing odors. Again depending on where one lives, you might have candles and sprays to "scent" your home, deodorants and soaps to make your body smell appealing, plastic lids to cover food or garbage. Communities tend to locate their "dumps" away from people's homes, although "toxic dumping" often reflects racial and socio-economic inequalities (Bullard 1990).

Noses and a sense of taste are deeply connected to cultural notions of "disgust." Feeling disgust means that one is repulsed by something; this is more serious than simply disliking something. Where scents come into play, we may be physically revolted by a noxious odor, such as a dead body with its rotting flesh. A feeling of disgust may be felt in the stomach, with waves of nausea, illustrating the connection to digestion.

What we find "disgusting" is both individual and culturally specific. One person may love the smell and taste of coffee, while for another, it may be quite literally sickening. Some people enjoy the scent of freshly-cut flowers, especially those with a rich bouquet such as roses, while others do not and develop coughing and sneezing reactions.

Different cultural contexts may also shape meanings of human body odor, with varying levels of acceptability. In some cultures, people shower or bathe daily; in others, much less often. Picking one's nose in many social settings is considered "disgusting," although it is fairly standard behavior among children the world over. The paper-products industry (e.g., Kleenex) is geared toward meeting the need to clean mucus and more tangible matter out of one's nose, replacing earlier technologies such as handkerchiefs.

Having a "good nose" often means that one is attuned to quality (or lack thereof) in food and wine production, perfume-making, and other endeavors involving scent. Having a nose might also mean that one is skilled at sleuthing and investigation; for example, he has a good nose for corruption. We also speak sometimes of something wrong or incongruent as smelling bad, as in "I smell trouble."

Like ears, noses are a site for body modification and adornment, with people in a variety of cultures sporting rings, bone ornaments, plates, sticks, and other items in their noses (often through their nostrils). Adornment may be for religious, secular, ethnic, sexual, or other reasons. Whatever the reason, modification is connected to social identity.

The nose is also subject to surgical intervention, with the erasure of noses that look "too ethnic" and their replacement with more "normative" noses. Gayle Kirschenbaum's film *My Nose* offers a humorous look at rhinoplasty in a Jewish family in the United States. Also called a "nose job," the procedure is popular among both women and men with financial resources in many countries.

Surgical intervention is often used to repair noses with congenital anomalies, such as orofacial clefts, or noses that have been damaged by injury. Because of their prominence on the face—they tend to stick out—they are easily hurt during contact sports (boxing and hockey are prime examples) or during accidents. Noses can also be damaged during physical fights and assaults. For example, a broken nose is a frequent consequence of domestic violence against women.

The nose is subject to conditions and diseases ranging from a "bloody nose" to polyps to rhinitis (or inflammation). During common colds and the flu, noses may "run" (i.e., leak mucus) or be "plugged up." Both circumstances can be uncomfortable and may affect breathing, eating, and sleeping. Treating these issues may involve "low-tech" procedures such as laying down with your head tilted back (for a bloody nose), to the use of over-the-counter nasal sprays, to use of pharmaceutical drugs.

Saddle nose is a condition in which the bridge of the nose collapses or is malformed, resulting in a concave or "saddle" shape. This can be caused by congenital syphilis (that is, syphilis present at birth), leprosy (also called Hansen's disease), and trauma, among other factors (Graper, Milne, and Stevens 1996). Where treatment is accessible, saddle nose can be repaired with rhinoplasty and/or antibiotics. However, in many settings in which it appears, particularly resulting from leprosy, treatment is not available.

Mouths, taste, and speech

Mouths are vital for human life. We eat, drink, talk, breathe, smile, frown, laugh, and kiss with our mouths. Comprised of lips, tongue, salivary glands, teeth, nerves, and muscles, our mouths are complex systems. When mouths do not work, whether for congenital reasons such as cleft lip and palate or for other reasons, our ability to survive may be diminished.

Mouths are especially interesting body parts, in that things go into our bodies through our mouths and also come out of our bodies through vomit, spit, and speech. When people talk incessantly, we sometimes describe them as "spewing words." A "mouthy" woman is one who is perceived to have oft-expressed opinions, or speech. The term is often used derogatorily, along with "She has a mouth on her."

The mouth is only one component of verbal speech; other parts include the larynx, vocal cords (or vocal folds), lungs, nose, nasal passages, and brain. Vibration of the vocal cords in the larynx, also sometimes called the "voice box," produces sound. However, in order to engage in speech, we must know how to form certain sounds or combinations of sounds, called utterances or words. We learn to speak through socialization in our families and communities of origin.

A significant part of our daily lives, eating and drinking happens (for most people) through the mouth. Thus, mouths (like noses) factor into taste, and whether we like something or not. Taste buds are receptors in our mouths, and they let us know which foods our body finds appealing. Taste is affected by smell—does the flavor invoke a good smell memory or a bad one?—and by experience.

In most cultures, kissing happens with the mouth. We can kiss somebody we care about—for example, a friend, a child, or a beloved aunt—on the cheek or hand, signifying affection. Or we can kiss somebody sexually, with our full lips and tongue. In this context, the mouth becomes a sexual organ, and its connection to kissing infuses its other functions with a hint of sexuality.

Mouths, and especially lips, are adorned and enhanced to sexualize appearance. Coloring the lips through shades and gloss is meant to mimic the bruised look of mouths after kissing, just as blush is meant to simulate post-conjugal flushing. Lip plumpers, Botox, and injections of collagen and silicon are designed to make lips and mouths look more inviting, and piercing of the lips and tongue may also hold sexual connotations.

The teeth are an important part of the mouth. They enable us to chew our food, and they also serve as a social marker. Good dental hygiene and preventive care is often limited to people with resources, for example dental insurance. People living in poverty or those with no access to dental care may have decaying teeth, may lose teeth prematurely, and may be at risk for infection and illnesses.

Keeping the teeth clean is big business, with multinationals such as Procter and Gamble and Colgate-Palmolive flooding the market with various kinds of toothpaste (e.g., whitening, cavity prevention, etc.), floss, dental sticks, and rinses. Straight, white teeth are perceived as a sign of overall health and prosperity, and in settings where people can afford advanced care, orthodontia (e.g., "braces") is common.

Our mouths are very busy body parts, which also renders them susceptible to contagion. We frequently touch our mouths, for example when we eat or pick food out of our teeth, and other people also touch our mouths, for example when we kiss or during infant feeding. Through eating, we are susceptible to food poisoning, and through breathing, to the inhalation of potentially toxic substances. Viral infections appear on and around our mouths as canker sores or cold sores (often caused by the Herpes virus). Halitosis (bad breath) can be a sign of underlying illness or infection. (Or, inadequate dental hygiene.)

Oral cancer, a potentially life-threatening illness, can affect the mouth, tongue, throat, neck, or any other part of the mouth including the lips. Although anyone can become affected with oral cancer, the risk is higher among people who smoke and/or drink alcohol, and also higher among people over age 40. Thus, oral cancer intersects with social judgments about "deviant" behaviors, which may affect prevention and treatment efforts.

A CLOSER LOOK

Botox

There is a range of cosmetic procedures designed to enhance appearance. Botox is somewhat unique, in that it involves the subcutaneous (under the skin) application of a bacterial toxin, *Clostridium botulinim*. This is the same bacterium that causes botulism, a dangerous and even life-threatening illness. So why would people want to inject a bacterial toxin into their bodies?

Botox has been approved by the United States Food and Drug Administration (FDA) for use in human beings to minimize wrinkles (e.g., frown lines around the mouth, crows' feet in the corners of the eyes). It has also been used "off label" to control severe underarm sweating, cervical dystonia (a neurological disorder that causes neck and shoulder contractions), and some vision problems.

Botox blocks nerves and weakens or paralyzes muscles, although not without risks.

For example, once injected the toxin can migrate through the body. Paralysis or weakening of muscles may be permanent, leading to lifelong facial disfigurement. Death is also a possibility. By 2008, there had been sixteen reported deaths in the U.S. from Botox (Public Citizen 2008).

Globally, the procedure is used widely by those with resources. Companies and Markets reports that the global Botox market could reach $2.9 billion by 2018. The overall facial aesthetics market is predicted to reach $4.7 billion by 2018, with the United States contributing $2 billion of the total.

What are we to make of this growing market for Botox? Given the efforts to keep toxins out of our bodies and communities, what factors can account for the deliberate injection of a bacterial toxin into our muscles and tissues?

Certainly, beauty standards have a great deal to do with use of Botox and other procedures, as women and men labor to create ideal body types. Feminist scholars such as Susan Bordo (1994) and Victoria Pitts-Taylor (2007) have drawn our attention to the complicated politics of age, gender, sexuality, and race that undergird use of cosmetic procedures.

However, it is also important to consider the growth of Big Pharma, direct-to-consumer advertising, and the emergence of new markets for capitalism. Communication scholar Susan Mello (2012) suggests that Botox has emerged as a "super cosmeceutical" in large part because consumers have been convinced through marketing that there are significant social and psychological risks of *not* using the product.

Here, we see the active construction of at-risk subjects who seek pharmaceutical intervention as a way to minimize their perceived risk. Eliminating wrinkles and lines, which are perceived and marketed as a kind of disfigurement, becomes a way to "save face" through normalization practices.

Facial recognition

Facial recognition technologies have emerged within the new, post-9/11 security apparatus. In a world ostensibly full of terrorists, governments and especially law enforcement agencies want to know how to identify potential assailants so as to prevent violence and also to punish perpetrators after a crime has been committed. In a world full of consumers, corporations

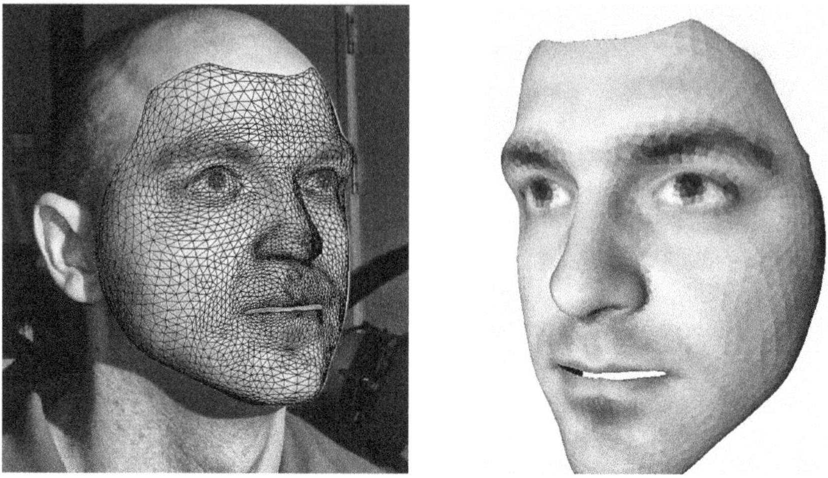

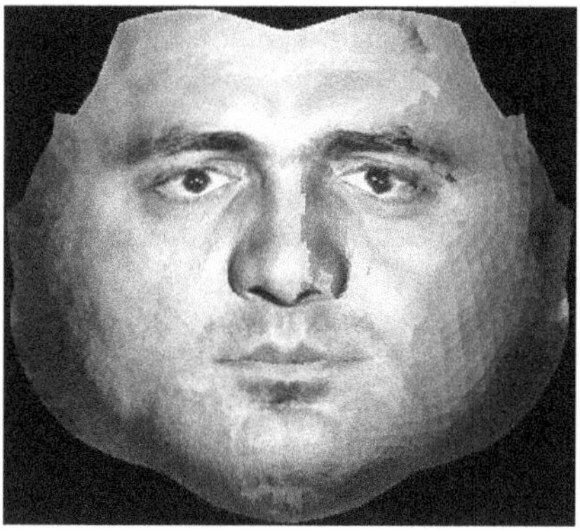

FIGURE 13.2 Facial recognition
J. R. Tena '*3D Face Modelling for 2D+3D Face Recognition*' PhD Thesis 2007, University of Surrey, UK.

want to know everything they can about who might be in the market for their goods.

Facial recognition relies on digital technology, and specifically a computer's ability to recognize a human face based on a series of criteria. Computers can be programmed to look for nodal points, which can offer distinctive data to identify an individual face (Li and Jain 2011). Human beings have 80 nodal points, measuring for example the distance between mouth and nose, or the depth of the eye sockets.

Once programmed, a computer can process visual images taken from a camera. Hidden surveillance cameras are often used to search for and process such images, especially at locations such as airports, train stations, and other potential targets. Li and Jain (2011: 1) state: "Face recognition has several advantages over other biometric modalities such as fingerprint and iris: besides being natural and nonintrusive, the most important advantage of face is that it can be captured at a distance and in a covert manner."

Of course, there is disagreement about what constitutes "nonintrusive" surveillance. From social media (e.g., Facebook's "tagging" feature) to employee oversight by companies, "covert" technologies raise ethical and legal questions. Do governments and corporations have the right to engage in surveillance of populations? If so, under what circumstances? What privacy issues ensue with the development of facial recognition databases? Is there such a thing as visual anonymity?

And perhaps most importantly, how might people protect themselves from facial surveillance, if they choose to do so? Sameer Padania (2012) offers a range of possibilities including tactical defenses in public, masks and other visual tricks, and technological solutions.

Conclusion

The face is a major site of social and species recognition, enabling various forms of human interaction. Through sociological notions such as "looking glass self" and "face work," we understand that faces and their multitude of expressions are important for relating to other people. A key component of these processes is the ability to imagine one's self, and then to orient behavior toward that self-perception.

Faces are the site of medical interventions, ranging from cosmetic surgery to face transplantation. These interventions, designed to "fix" faces or to improve them toward greater normativity, reflect and perpetuate social inequalities of difference. Certain faces are interpreted as "disfigured" and interventions are used to treat the disfigurement.

The face's major features—eyes, ears, nose, and mouth—are sites for multiple meanings and functions. In different ways, each of these body parts is also central to social interaction through sensory engagement and communication with other human beings, with nonhuman animals, and with the material world around us.

See also

Respiration; Endocrine system and metabolism; Digestion; Skin

Key terms

Looking-glass self, stigma, total institutions, immunosuppressant drugs, disfigurement, observation, ethnography, visual sociology, Deaf culture

Discussion questions

1. What does it mean for someone to "lose face" in your particular social group? What are strategies for "saving face" in your social group?
2. Forecasting into the future, what do you anticipate is on the horizon for face transplantation surgeries? How does the slipperiness of the category of "disfigurement" enable some "normal" facial features to be classified as disfigurement?
3. What smells good and bad to you? How might these be socially constructed and rooted in your history?
4. Should children who are unable to hear be fitted with cochlear implants? Why or why not?
5. Governments and corporations are increasingly using facial recognition technology to identify people as suspects, consumers, or intruders. What are the benefits and costs of this technology, and to whom? Is facial recognition technology ethical?

chapter 14

Skin

Introduction

Brazilian artist Angelica Dass's art records and catalogs human variations in skin tone and matches them to the Pantone color schema. Pantone, headquartered in New Jersey in the United States but with offices and facilities worldwide, is considered a leading authority on color classification and color matching. Dass's project, entitled "Humanae," is astonishing in the way it explodes the notion of "flesh-colored" and illustrates the limitless variations in skin tone between each individual. There is no singular meaning of the color "flesh."

Our largest and most visible organ, skin provides protection from the elements and is a source of physical sensation, or touch, as the skin harbors several types of nerve endings. Humans have approximately 22 square feet of skin, which comprises about 16 percent of our body weight. Skin, or more precisely the epidermis, is self-regenerating as we slough off about one million skin cells a day; a typical skin cell lasts about two weeks. Where do these cells go? They become dirt and debris that fills your home, office, and classroom, for dust is mostly composed of dead skin cells.

Skin can be of different colors or shades, called *pigmentation*. The more melanin, a pigment produced by melanocyte cells in the skin, a person has, the darker their skin will be. Melanin protects the body by absorbing solar radiation, and in hotter, brighter southern climate zones, darker skin is understood scientifically as an adaptation. And yet, darker skin is understood socially and politically as a marker of inequality, as is aged and wrinkled skin, particularly for women in many cultures.

FIGURE 14.1 Pantone skin chart
© "Humanæ" – Angelica Dass

In this chapter, we explore the ways in which human skin is a social signifier. A person's skin color, tone, complexion, and body modifications are used as visual information to locate people within systems of social stratification. *Social stratification* is the division of people in a given society into unequal levels according to relative power, property, and prestige. Examples of stratification are everywhere in our social interactions and social groups—both locally and globally.

In educational institutions, individuals are stratified based on their academic performance measured by their grades whereby those with higher GPAs are treated with more prestige, given more privileges, and treated with more respect. In capitalist economies, people are stratified based on their relative wealth, occupational category, and material possessions. Many different cultures and countries stratify individuals based on their perceived race and/or ethnicity as identified or interpreted via skin color.

These social dynamics have resulted in a number of practices focused on evaluating and altering skin, particularly (but not only) in developed nations. These include skin lightening and darkening through chemicals, discriminatory practices such as colorism and racial profiling, and anti-aging interventions ranging from moisturizing cream to dermabrasion to chemical peels.

As a form of self-expression or group membership, individuals also may adorn their skin with temporary or permanent marks that change the pigmentation of the skin. Tattooing has been in existence since prehistoric times, and its cultural history illustrates how definitions of deviance and social acceptability change through generations.

Race and "color"

Sociologists distinguish between differences associated with race and differences associated with ethnicity. Race is considered by most to be associated with physical characteristics or appearance and is often in the eye of the beholder. Bodily characteristics are used to decipher someone's race—the shape of the eyes, texture of the hair, color of the skin, and body size—often without explicit corroboration from the individual.

Building on work of abolitionist Frederick Douglass, in the early 1900s activist, sociologist, and historian W. E. B. Dubois developed the concept of the color line. He stated: "the problem of the 20th century is the problem of the color line—the relation of the darker to the lighter races of men in Asia and Africa, in America and the islands of the sea" (Dubois 1903). The *color line* has come to mean social or political restriction or distinction based on differences of skin pigmentation.

Civil rights violations such as slavery, segregation, and voting are obvious examples of how the color line separates people based on perceptions of race as indicated by skin color. Even though legislation can create a distribution of rights that challenges the color line, beliefs about race persist along with racist practices. For example, a number of scholars have examined the U.S. prison-industrial complex and its inherent racism (Gilmore 2007).

The social construction of race is the process by which people come to define a group as a race based on physical characteristics and historical, cultural, and economic factors. Ethnic groups are comprised of individuals who share a set of cultural traits, including food, celebrations, clothing, or country of origin. Race and ethnicity are considered to be an *ascribed status*, which means they are attributed at birth and typically do not change over the lifetime of an individual.

Social scientists and others use the word *prejudice* to describe negative attitudes toward an entire category of people, often an ethnic or racial minority. Ironically, when prejudice is expressed or experienced, it is often without confirmation that an individual is of the perceived race. Very often, skin tone along with hairstyle, clothing choices, or facial attributes, are used to label another's race.

Skin tone might trigger feelings of prejudice, but presumptions about ethnicity are many times inaccurate. For example, in New York City, based on phenotypic characteristics including skin color, individuals who are Puerto Rican can be perceived to be African American, or Italians can be read as Latinos. So prejudice can be expressed toward individuals who may not be members of the maligned racial group. Prejudice can result from *ethnocentrism*, which is the tendency to assume that one's own culture and way of life represent the norm and/or are superior to others.

Coined by feminist theorist Peggy McIntosh (1988), *white privilege* describes the advantages and immunities shared by those who are categorized as "white" in a given culture. Importantly, white privilege is invisible to most whites because of the taken-for-granted notion that white is neutral, or non-raced. Since white privilege is often unacknowledged by white people, even within activist causes, whiteness complicates the notion of "equal rights."

Indigenous rights scholar and activist Aileen Moreton-Robinson (2003) explores how throughout the national Australian project, white women benefitted from decades of colonialism of indigenous peoples in Australia. White feminist Australians were fighting for equality with men, so clearly aligned with white men's rights, and thus "Whiteness [became] so pervasive as an invisible norm that race, as difference, still belongs only to women who are not white in Australian feminism" (2003: 110).

Racism is a form of prejudice holding the belief that one race is supreme and all others are innately inferior. When racism prevails in a society, members of subordinate groups experience prejudice, discrimination, exploitation, and violence. Racism may be upheld through institutional and state practices, such as border and immigration controls.

Discrimination is the act of denying opportunities and equal rights to individuals and groups because of prejudice or other arbitrary reasons. The term "glass ceiling" refers in some contexts to an invisible barrier that blocks promotion of a qualified individual in a work environment because of the individual's gender, race, ethnicity, or sexual orientation.

One very clear example of discrimination is the case of racial profiling which takes many forms. As we examine below, in the case of automobile or airplane transportation, people who are not seen as "white" face challenges that are deeply consequential for their freedom of movement.

A CLOSER LOOK

Racial profiling

Driving while black

A transparent example of how discrimination based on skin color operates in contemporary American society would be the use of racial profiling, which in part is sustained by racialized transportation practices that "police" mobility. While skin does transmit information to those enacting racial profiling, gender, clothing, hair, and accent also come to bear on the social interaction.

Security guards, police officers, shop owners, and border patrol agents are charged with maintaining social order and operating within the norms of civil society. But sometimes, those who are charged with managing social spaces treat individuals as suspects based solely on racial or ethnic characteristics instead of using evidence-based judgment of criminal behavior. We label this behavior racial profiling, or the use of racial characteristics as the basis for criminal designation (Glover 2009).

One of the most compelling cases of racial profiling is that of the New Jersey Turnpike. Black motorists were pulled over by the New Jersey State Police at disproportionate rates to all other drivers. In 1998, a van carrying four unarmed men of color was stopped on the turnpike for speeding. The van started to roll backwards after it stopped, and the two officers opened fire on the van, injuring three of the men.

This incident led to a federal investigation in which police officers admitted that their supervisors directed them to pull over drivers who appeared black. In fact, black drivers were five times more likely to be pulled over than other drivers, and 15 times more likely to be arrested.

For 10 years, the United States Department of Justice, through a consent decree, assumed oversight of the training and evaluation practices of the New Jersey State Police.

Airport security and surveillance

Post 9/11, airport security intensified considerably with increased surveillance, new technologies, labor-intensive procedures, and elevated passenger frustration. When passengers travel through airports, they must be screened prior to boarding in the "security line" by airport security personnel.

In the United States, these people work for the TSA, or Transportation Security Administration, and in Canada, they work for the Canadian Air Transport Security Authority (CATSA). The Australian Government has implemented a program called TravelSECURE, which provides information and services. In other countries, security services may be contracted out; for example, in Finland security is provided by the company SOL Security Service Oy. In India, a paramilitary unit called the Central Industrial Security Force maintains airport security.

With all of these different providers, there is little international standardization of protocols and procedures. Even in different airports in the U.S., where we live, security procedures may vary, despite the 2001 Aviation and Transportation Security Act. What also may vary are the experiences, prejudices, and cultural understandings that airport security personnel bring to their jobs. This includes ideas about race and citizenship.

There have been numerous incidents in the United States, for example, in which people of color, and especially people of Middle Eastern descent, have been racially profiled. That is, they have been detained or singled out not because of "suspect" behavior but because of obvious or perceived racial characteristics. Operating with a national security lens, some people are presumed on the basis of race to be potential "terrorists."

In 2012, the Transportation Security Administration came under fire for its "behavior detection program" at Boston's Logan International Airport. Originally intended to detect suspicious behavior that might signal terrorism, the program quickly began to profile *all* racial minorities including Hispanics bound for Miami and blacks wearing "suspicious" clothing, such as baseball caps turned backward. Members of minorities eventually made up an "overwhelming" number of cases referred to the Massachusetts State Police, triggering an investigation of TSA's practices.

Here, we see skin color standing in for not only race, but also for "suspicious" behavior and perceived criminality, suggesting at least one way in which human skin is enmeshed in culture and politics.

Cross-cultural understandings of racial difference

It is important to note that although we have been discussing the United States, meanings of race vary cross-culturally. White, black, and brown signify different identities in different contexts, and these are further saturated with meanings about ethnicity, class, gender, and geographic origin.

For example, historian Warwick Anderson (2003) has chronicled the role of race, and particularly indigeneity, in the formation of Australian nationalism. In a study of south-eastern Brazil, sociologist France Winddance Twine (1998) discussed the differences between local meanings of race and racism among Afro- and Euro-Brazilians and the meanings used by researchers, noting that meanings of "black," for example, are quite different in Brazil than in the U.S. and Europe.

Japan, on the other hand, has been termed a "monoracial" culture even though people of other races inhabit the country. There are reports of racism so prevalent that the country has been investigated by the United Nations (more than once) for discriminatory practices. The Yamato are the native dominant ethnic group of Japan, while the Ainu—originating from the Okhotsk Sea region near Russia and viewed as ethnically not Japanese—are recognized formally as an indigenous people.

Preferences and practices to achieve whiteness predate colonialism, while at the same time colonialism reinforced and amplified the structural preference for whiteness. Indeed, many have argued that the "desire for white skin is universal" (Isa and Kramer 2003). This desire is not just a personal or group preference based on aesthetic beliefs but rather deeply linked to access to power and prestige.

Scholars in Asian cultures, for example, have demonstrated how having white skin affects women's marital prospects, employment, social status, and earning potential (Ashikari 2003). A team of Canadian, Japanese, and American researchers analyzed print advertisements for skin whitening and lightening products in India, Hong Kong, Japan, and Korea and argued that white skin is a form of cultural capital in Asian society (Li and others 2008). Whiteness is pursued to signify belonging to upper echelons of social orders.

In her fascinating book *Pure Beauty: Judging Race in Japanese American Beauty Pageants*, sociologist Rebecca King-O'Riain explores the relationship between race, gender, and sexuality, and the many ways that "racial eligibility rules" or the rules of who can participate in the contests, structure pageant participation. Herself a Japanese-American former beauty pageant contestant, King-O'Riain goes beyond auto-ethnography and employs a methodology of documentary and archival research, interviews, and participant observation of young women in beauty contests in San Francisco, Los Angeles, Seattle, and Honolulu.

Throughout her research, King-O'Riain explains how Japanese American beauty queens are supposed to serve as "symbolic representations" of their community—the racialized standards of what it is to be Japanese American. The rules are constructed for Japanese Americans in the United States but rely on historical and culturally specific understandings of Japanese identity and racial purity in times of multi-racial mixings. Skin tone is a major factor in these understandings.

Despite considerable cross-cultural differences in how race and racism are identified, in almost all nations, race is linked to skin color, which in turn is linked to cultural understandings of blood, kinship, place, and genetic identity. Racism is almost always linked to structural inequalities.

Colorism is a belief in the superiority of lighter skinned people over darker skinned people, which results in documented outcomes in distribution and access of resources in areas such as income, education, housing, and interpersonal relationships. Although colorism is sometimes synonymous with racism, it is also commonly practiced among people of the same racial/ethnic group.

Joanne Rondilla and Paul Spickard (2007) examine how color hierarchies operate in and among different Asian cultures, including Laotian, Japanese, Vietnamese, and Korean. Their historical and ethnographic research reveals: "The feature of Asian American colorism that immediately stands out as being somewhat different from the Black and Latino version is that Asian American colorism seems not to be mainly about Whiteness at its point of origin. For Asian Americans, color hierarchies seem to be as deeply rooted in old-country class distinctions—in the desire to look like upper-class Asians who did not have to do body work out in the sun—as in U.S.-generated desires to look like White people." (15)

In North America, colorism has been examined in several large-scale survey studies. Sociologist Margaret Hunter's (2002) analysis of two U.S. national databases demonstrated how lighter skin tone benefited African-American and Mexican-American women in income, educational attainment, and spousal status. According to research conducted in Mexico by sociologist Andrés Villarreal (2010), Mexicans with darker skin tones face discrimination in the labor market. Darker skin tone ("looking Mexican") increases the vulnerability of Latinos/Latinas in the American Southwest to surveillance and arrest, particularly in states such as Arizona with harsh anti-immigration policies and militarized intervention at the U.S./Mexico border (Hernandez 2010).

Evelyn Nakano Glenn, a sociologist, calls skin tone an element of symbolic capital whereby the lighter the skin, the more access to power, opportunity, and prestige. A multi-billion dollar transnational skin-lightening marketplace helps individuals to lighten and brighten dark and aging skin. As Glenn (2008: 298) argues: "The yearning for lightness evident in the widespread and growing use of skin bleaching around the globe can rightfully be seen as a legacy of colonialism, a manifestation of 'false consciousness,' and the internalization of 'white is right' values by people of color, especially women."

Given this colorism that creates disadvantages for those with darker complexions, it is not surprising that there is a market for skin lightening or bleaching products. However, lightening does not equal health. A team of dermatologists (Olumide et al. 2008) studying a range of countries on the African continent found the active ingredients in many of these products to be hydroquinone, mercury, and corticosteroids, with severe side effects including nephropathy, steroid addiction syndrome, predisposition to infections, and death.

Sociologists Phillip Vannini and Aaron M. McCright (2004) also examine skin color, in particular the artificially tanned body. Based on qualitative interviews conducted with 40 people, they analyze the practices of those who regularly tan themselves at salons that offer tanning beds and lamps. Many do so to achieve darker skin tone, which is often read pejoratively within the context of racial difference and inequality, but paradoxically, as a sign in some cultural settings of youth, health, sexuality, and beauty.

For some people who identify as white, or lighter skinned, achieving a tan is a way to make the body more beautiful. Comparable to the rise in cosmetic surgical procedures, the growth of the tanning industry in the United States has been astronomical since the first commercial beds appeared in 1974. Even though the American Medical Association and American Association of Dermatology, as well as federal agencies like the FDA, assert that tanning is physically risky behavior, tanning is a billion dollar business. In the UK, tanning among youth ages 11–17 is "widespread," with 50 percent of girls from larger cities reporting they had used a tanning bed.

Peer-reviewed medical research finds that prolonged exposure to solar UVA and UVB radiation places one at risk for all three types of skin cancer. Even artificial tanning is potentially harmful—there is no such thing as a "safe" tan. But medical discourses compete with cultural discourses, what Vannini and McCright refer to as the "seduction frame." The seduction frame, fueled by both the tanning industry and celebrity culture, explicitly plays down the risks of tanning. Pushing these risks aside, the seduction frame advances the notion that tanned skin heightens beauty, makes you look more fit, hides wrinkles, and improves overall physical and mental health.

The authors argue that the enticing image of a bronze body is appealing because it offers semiotic power, or symbolic value. This form of cultural capital demonstrates that someone has the leisure time and excess income to purchase a tan, travel to a beach resort, or sunbathe presumably in a nonworking context—in fact, in some cultures we distinguish between a "farmer's tan" and a salon tan. Rather than signifying damage and illness, tanning aficionados see their bodies as seductive and empowered, and their "healthy glow" as a status marker connoting membership in the "leisure" class.

Here we can see how the body is both a medium and a symbol. In this case, for white or lighter-skinned people achieving darker skin is viewed as a sign of beauty and self-cultivation. Unlike racialized discourses that construct those with biologically darker skin as inferior and devalued, artificially tanned bodies are viewed in some contexts as sexy and empowering. The cultural politics of human skin underscore how the body is at once a biological and social organism.

Skin as we grow and age

Skin has been described with socially value-laden words—positively we can state that skin is supple, bright, taut, clear, or glowing; negatively we might describe skin as wrinkled, lined, dull, leathery, sagging, blotchy, or sallow. Just as social stratification about race is expressed in colorism rampant in many cultures, a social preference for youth is expressed in the ways we manage aging skin.

As skin ages, collagen and elastin break down, making it more difficult for skin to heal and depleting elasticity. Wrinkles become more prevalent and visible as we age. Medical procedures and health and beauty aids conduct a brisk trade in "treating" wrinkles.

Additionally, while certainly not universally true, there is a strong social preference for slender bodies in the West – a preference that is rapidly being exported across the globe. As fat studies scholars Rothblum and Solovay argue:

> Weight-based discrimination is a cradle-to-grave phenomenon. Fat people are officially barred from adopting babies in China. In Britain, health clinics may refuse in vitro fertilization to fat women, and the British Fertility Society has recommended a general ban. In the United States, public health departments advertise that parents should prevent childhood "obesity" . . . Girls as young as five fear gaining weight. At age ten, fat children are chosen last as friends by peers (Rothblum and Solovay 2009: xix).

Being "comfortable in your skin" is an expression used in some cultures to suggest self-confidence or the level of ease an individual feels with themselves. However, given the social messages we receive about our aging, stretching, or splotched skin, it is very challenging to maintain self-confidence. Those with the means to transform their skin may choose to engage in body projects that achieve more socially acceptable appearance. Furthermore, there are real material benefits in terms of employment, interpersonal relationships, and fertility services related to how one's skin appears.

Touted as procedures that erase the effects of time or weight gain, a variety of aesthetic or cosmetic interventions include:

Lipoplasty: A modification of body contours by removal of excess adipose (fatty) tissue, usually on the stomach, thighs or buttocks.

Hyaluronic acid injections: A type of filler injected directly into the tissues to add volume to the skin, usually the cheeks and lips.

Botox: Small injections of a bacterium to paralyze facial muscles, which smoothes out wrinkles most commonly on the forehead and around the eyes (see previous chapter).

Lifting skin surgically: Skin around the eyes, neck, forehead, and eyelids is cut and sutured to pull the skin more tightly across the face and neck. The International Society of Aesthetic Plastic Surgery compiles statistics on the global prevalence of plastic surgery. The top five countries for plastic surgeries and plastic surgeons are, in order from most prevalent, the United States, Brazil, China, Japan and Mexico. Notably, the U.S. and Brazil are very close in number.

Given the discrimination and stigma associated with fatness, it is not surprising that in 2011, lipoplasty was the most common form of plastic surgery followed closely by breast augmentation. Coming in third was blepharoplasty, or eyelid surgery, to reduce sagging eyelids or construct a double eyelid in the case of Asiatic eyes.

Asian blepharoplasty is popular in Taiwan and South Korea as well as the United States among Asian Americans, and it has been widely criticized. Author and poet David Mura has stated: "It's evidence of internalized racism. It really indicates something about the way in which Asians in America are indoctrinated by white standards of beauty. They feel less beautiful than those who fit the Caucasian standard of beauty."

Scars

Anyone who has ever cut herself with a knife, scissors, or other sharp object or burned himself on a hot skillet knows that a scar forms over the cut. Scars are the skin's way of healing. They are made of the same protein (collagen) as our regular undamaged skin, but they are more fibrous, which is why they appear different.

Some scars are badges of honor for children, who show them off to their friends with an ever-embellished story. Even adults find meaning in scars, as visible evidence of a life well (or dangerously) lived. ("Remember that time when . . .?") One of us, for example, has had two cesarean sections, and the scars are positive, tangible reminders of childbirth and mothering. The other has scars from orthopedic surgeries from injuries sustained in collegiate athletics, reminders of past lives and abilities.

Scars are also a form of body modification, called *scarification*. Here, scars are scratched, burned, etched, or cut into people's bodies for social, religious, aesthetic, or cultural reasons. Scarification is also used in some cultures as a medical treatment for various ailments. Scars of a similar type may bond people into communities or open up the possibility for human connection when scars are visible.

In the case of self-injury, sometimes called cutting, sociologists Patricia and Peter Adler (2011) explore through some 150 interviews the non-suicidal destruction of one's own body tissue. Their ten-year longitudinal study finds that self-injury to the skin can be an expression of group membership whereby the scars serve as evidence of commonality among a peer group and a form of harm reduction. That is, self-cutting prevents other, more dangerous practices, such as suicide.

Other scars have different, more negative meanings: scars formed from beatings and assaults; the scars that may criss-cross somebody's wrist after a failed suicide attempt; the scars left by weapons in wars, ethnic conflicts, and massacres; surgical scars that mark accidents, amputations, and other traumatic events.

It is not surprising that the word "scar" has multiple uses in the English language. Etymologically, the word derives from the Old French *escharre*, meaning hearth or fireplace; thus, scarring was related to burning. In medicine, the term was used historically to refer to scabbing, as over a wound. In both lay and professional use today, people who have been hurt or traumatized are described psychologically as "scarred," which sometimes equates to having a damaged psyche.

Where there are scars—physical and psychological—there have also been attempts to erase them. Psychotherapy has been used to repair damaged psyches, and various treatments have been used to erase physical scars. For example, surgical removal, laser removal, and steroid injections are frequent treatments for scars, as are creams, gels, and ointments. Vitamin E is often recommended.

New treatments may also be on the horizon. In the 1980s, fetal surgeons engaged in clinical work discovered an interesting and promising sideline—fetal wound healing (Casper 1998). Fetuses that were operated on *in utero* and subsequently delivered were found to have no scarring. This quite extraordinary finding suggested that something in the amniotic environment or fetal cells was conducive to healing scars quickly, and a research agenda was born.

It did not take long for these findings to work their way into various commercial applications. Fetal tissue engineering and fetal stem cell research is increasingly linked to cosmetic and reconstructive surgery, regeneration of body parts, treatment of fetal and infant congenital anomalies, military medicine, and the cosmetic industry.

For example, the McKay Orthopaedic Research Lab at the University of Pennsylvania is investigating fetal tissue for use in tendon regeneration. Similarly, Traci Wilgus (2007) has written about fetal wound healing in the context of ostomy management. In 2009, a San Francisco-based company called Neocutis, Inc. was heavily criticized for using cells from an aborted fetus in skin care products (Marietta 2010).

The "brave new world" of fetal tissue engineering thus raises questions about appropriate uses of human biological materials. Scarring and various interventions for removing scars animate many of the issues we have been discussing throughout this book: that bodies have multiple meanings and are situated in varied social, political, cultural, and economic contexts.

A CLOSER LOOK

Tattoos

New York Knicks basketball player JR Smith has called his 70-plus tattoos a lifestyle. He is not alone in using voluntary body alteration as a medium to express himself beyond a temporary application of cosmetics, jewelry, or clothing.

Human skin, in many ways a canvas of self-expression, is also marked as a way of demonstrating cultural membership in familial, social, institutional, or ethnic groups. Tattoos have signified genealogy, tribal membership, developmental maturity, and gang affiliation, and they also may indicate experience in jails or prisons, military service, or religious association.

As a means of understanding the renaissance of tattooing in North America and Western Europe, cultural scholar Juniper Ellis (2008) explores the socio-cultural, aesthetic, political, and literary significance of tattooing in New Zealand, Hawaii, Tonga, Samoa, the Marquesas, and Tahiti. In many of these cultures, tattoos are forms of expression that connect the physical to the metaphysical in spiritual practices. Mobile Pacific Islanders and those who have traveled to these geographies, in addition to cultural exchanges precipitated by colonization and militarization, facilitate the movement of tattooing and tattoos from indigenous to non-indigenous populations. As Ellis explains, nearly 40 percent of all Americans under 35 years of age who live in the United States are tattooed.

In addition to the way skin modifications and adornments have traveled through time and across continents, the shifting social acceptance of tattooing demonstrates how social norms also evolve and shift. Social stigma previously associated with acts of tattooing in some cultures has been reframed as "popular" or even part of an artistic canon (Kosut 2006). Indeed, according to sociologist Clinton Sanders (2008): "In addition to being incorporated into the lucrative world of popular culture, in the latter part of the 20th century tattooing also became more firmly situated in the world of 'serious' art."

However, the integration of tattooing into legitimate social institutions does not create universal acceptance. Sanders adds: "Employers still reject tattooed job applicants and parents still bemoan their children's decisions to get tattoos . . . the medical, psycho-therapeutic and public health industries continue to cast body alteration as a dangerous practice or indicative of underlying pathologies" (xiii).

Among subcultures the acceptability of tattooing varies: tattoos can be socially celebrated, even expected, or professionally reviled.

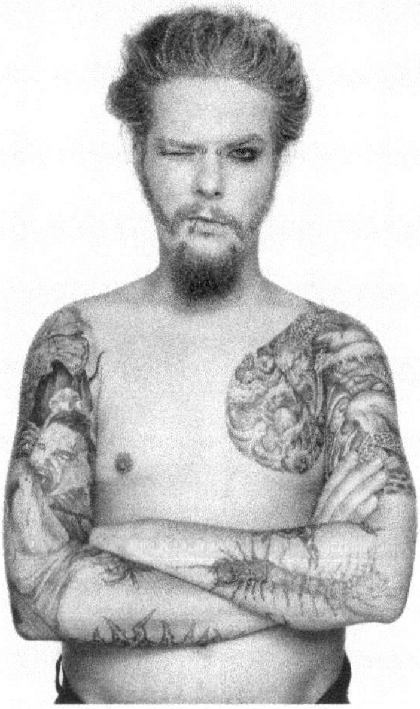

FIGURE 14.2 Tattoos
© Shutterstock

Conclusion

Unlike our other bodily organs, skin is almost always visible to ourselves and others. It protects and contains us, providing a fleshy, tangible barrier between our own bodies and the rest of the world. However, we may modify our skin in a multitude of ways, in order to fit into social systems and communities or to express ourselves given our unique cultural location.

There is often a relationship between what we do to our skin (e.g., tattoos, scarification, anti-aging treatments) and how we *feel* in our skin—both physiologically and psychologically—as human beings who are part of multiple social worlds in which our bodies are evaluated for their social worth. Skin always interacts with multiple other bodily signs and symbols (e.g., gender, facial structure, hair, clothing, social circumstances, geography) in that people are located—or locate themselves—within systems of social stratification.

And like other body parts, skin holds multiple meanings and is both understood and treated within institutional settings such as clinical medicine. It also can be mapped along axes of social inequality, with skin color serving as a marker of race, ethnicity, and value.

See also

Cells and tissues; Organs; Face; Hair

Key terms

Pigmentation, social stratification, ascribed status, prejudice, the color line, white privilege, discrimination, racism, colorism, racial profiling, ethnocentrism, scarification

Discussion questions

1 Think about the adjective "flesh-colored," used historically by Crayola to describe the crayons renamed in 1962 to "peach." What image comes to mind when thinking of the color flesh? What color is *your* skin? Does it match the implied and socially agreed upon color of human flesh? How might this be an example of white privilege?

2 Racial profiling of individuals perceived to be of Middle Eastern descent has increased significantly at many international airports since September 11, 2001. However, this racial profiling has not proven to eliminate threats of individuals and possibly creates a diversion from other ways of mitigating danger. In what ways is racial profiling practiced and how does skin figure prominently in racial profiling? What are the pitfalls of the practice?
3 Getting a spray-tan, which runs about US$45 a session, is a popular body project to attain a "healthy glow." What are some of the larger social forces operating when an individual acquires a spray tan? How do we react when someone has a tan? How does that reaction change when we realize they paid for it at the mall rather than acquired it "naturally" on vacation?
4 What scars do you have on your body? What kind of story do they tell about your life? Is it a story you would share with other people? If you could remove the scars using a cream made of fetal tissue cells, would you do so? Why or why not? Would you use any other technology for scar removal?
5 How do types of surgical procedures on the skin indicate social preferences for particular types of bodies? In other words, what types of bodies are culturally esteemed in different cultures? Think about the range of body types and skin types. Does the exporting of the "Western ideal body" threaten to transform bodies, and in particular skin, to something more homogeneous?

chapter 15

Hair

Introduction

Along with being warm-blooded, giving birth to live young, and producing milk from mammary glands, growing hair is a biological characteristic that defines mammals. Yet among mammalian species, we would venture, it is humans who most fret about, adjust, and invent products for our scalp, body, facial, and pubic hair. True, our nonhuman primate cousins spend a great deal of time socially bonding through extensive grooming rituals, but we doubt they "stress out" over a bad hair day or ever consider donating, selling, and buying their fellow primates' hair. Have you ever seen a gorilla in a wig?

Physiologically, hair serves numerous purposes: keeping humans warm, protecting us from the sun's rays, enhancing our sense of touch, and in the case of nostril hair, filtering out particles from entry into our respiratory system. We have approximately 5 million hair follicles at birth and lose on average 80 strands of hair per day (Paus and Cotsarelis 1999). Even though it is not always visible, hair covers most of the human body, except for the lips, palms, and soles of our feet.

In this chapter, we move beyond biological understandings of human hair to consider our "crowning glory" as rife with symbolism and status. From Jazz Age bobs to the 1950s bouffant, long-haired hippies to mohawked punk rockers, mullets to rat tails, and dreadlocks to "locks of love," hair says as much about cultures and social meanings as it does about the person sporting it. Hair also demonstrates elective or forced group membership, as in the case of shaving hair to mark a group affiliation in the case of bald "skin heads," or institutionally to control a group of people, as in the military or concentration camps. We use hair to signify status on many different levels of human interaction and group organization.

We begin with a discussion of the ways that meanings of race and gender are entangled with hair, including an exploration of the hair trade. An experiment that interprets the social and interpersonal consequences of not shaving or shaving gets a closer look. Baldness, as part of the natural life course or brought on through the treatment of disease, is discussed as being associated with high degrees of social stigma and subsequent social management, but also a medium for solidarity and social activism.

We also explore "manscaping" and how it has increased in popularity among some men. Finally, we conclude by explaining how we come to coif and groom ourselves through ritualized patterns that satisfy or defy social norms and cultural standards.

Gender, race, and hair

Even before we are aware that someone is looking at us, our hair is used to provide social information to others. When identifying a person's sex or race (or social class), hairstyle (length, cut, color, shape, texture) is often used as a clue or marker (Brebner 2009). As sociologist Anthony Synnott (1987) suggests, because hair is personal and public as well as malleable, it occupies a cultural space as integral to both individual identity *and* group identity. Indeed, throughout history and spanning diverse cultures, hair has been written about in poetry, music, and romance novels as synonymous with sexual prowess and desirability, virility and health, and youth and vigor.

Moreover, by the particular locations of hair growth and removal on male and female bodies, we affirm that we are largely *sexually dimorphic* (meaning we are a species of two distinct sexes). For example, in many cultures, social norms dictate that facial hair belongs on males and that females should have long, flowing locks but be virtually hairless everywhere else. Women who shave their heads, such as Irish musician Sinead O'Connor, often do so to make a feminist or other social justice statement. Among Muslims, Orthodox Jews, Sikhs, and the Amish, for example, beards on men are mandatory, and to cut another man's beard is a serious offense.

Sociologist Rose Weitz, through in-depth interviews of a racially and generationally diverse group, studied girls' and women's relationships to their hair as well as the social consequences of styling choices. Weitz states: "whenever we cut our hair or grow it long, cover the gray or leave it alone, dye it blonde or dye it turquoise, curl it or straighten it, we decide what image we want to present in the world. And the world responds in kind, deciding who we are and how to treat us based in part on what our hair looks like" (2004: 219).

Hair is thus both a creative enterprise for humans as well as a visible way of signifying certain social statuses and identity cues. So managing our locks is not just a personal idiosyncratic expression of self; it also reflects the internalization of social norms and values about the presentation of self (Goffman 1959) and our place in social worlds. Hair is also a crucial site for challenging these same norms.

A CLOSER LOOK

To shave or not to shave

The gendered presentation or removal of body hair has significant meaning for different cultures and even different generations or cohorts. Psychologist Breanne Fahs conducted an experiment with her college students. Men were instructed to shave off all their body hair and women were told to grow their body hair for 10 weeks; both groups kept detailed journals of their social interactions (Fahs 2012). As masculinity and femininity are in many ways supported by the management of body hair, Fahs expected to see some differences in how individuals reacted to the men and women. The experiment, voluntary among her undergraduates, was conducted over several semesters with different classes.

Fahs explains that body hair "carries with it a host of ideologies about 'appropriate' gender, normative behavior, conformity, and the (lack of) acceptability of hairy or hairless bodies" (2012: 486). However, despite her hypothesis of the reaction to the experiment regarding behavior of men and women's body hair management, the results were nonetheless startling. Parents, friends, and partners expressed shock, anger, and shame regarding the hair-grooming experimental subjects. Furthermore, the reaction was different for individuals based on race, sexual orientation, and social class—as Ana's comment (p.494) below illustrates:

> When I compared my hair to the hair of the other girls in class, there was an obvious difference. My hair grew in thick and coarse. The other Latina women in the class understand that the white girls had it easier because their hair was thinner. I felt like people would think I was a "dirty Mexican" because of the hair, that I was doing something nasty, and people would connect my body hair to my being lesbian or Mexican.

The experiment had a consciousness-raising effect on the men, some of who were able to tap into their empathetic selves. One subject stated: "I found myself spending about 10–15 minutes shaving, thinking I did a good job, just to step out of the shower, put on my glasses, and realize that I did not do so well. I would then stop and go back and try to do it again, which was very frustrating. I slowly began to realize what women go through just to fit in" (496).

As Fahs's pedagogical experiment shows, hair requires a level of personal and cultural management to make it socially acceptable: cutting, shaving, dyeing, weaving, straightening, growing, styling, and so on. In short, "natural" human hair growth and color, particularly but not only in industrialized countries, is often not the norm.

We all have an ethnicity, so why don't we all have "ethnic hair"?

Almost all drug stores, pharmacies, and chemists have a hair care aisle, a section stocked with shampoos and conditioners, styling products and hair dyes, combs and brushes, and accessories like headbands and barrettes. Typically off to the side, at the end of the aisle or on the bottom shelf, is the Ethnic Hair Care section. In the United States, "ethnic hair" most often refers to African American hair. Products in the ethnic hair care section often include relaxers, curl activator, texturizers, and straighteners.

Kinky, curly, nappy, and wooly are all adjectives used to describe African American hair, although not all African Americans have this type of hair. Furthermore, many other racial/ethnic groups have hair that could be described in this way. However, because we live in a highly stratified and racialized world, biological differences are used to create categories of people. The very creation of a separate section for those with "ethnic hair" suggests that there are other individuals with hair that is not "ethnic" and is, rather, "normal" hair.

For example, in 2009, African American comedian Chris Rock produced and narrated the documentary *Good Hair* in response to his then three-year old daughter asking him: "Daddy, how come I don't have good hair?" The film traces the historical styling of African American women's hair and the way relaxers are used to straighten hair, often at great discomfort and expense to girls and women.

A political response to these norms emerged in the 1960s with the Black Power movement in the United States and United Kingdom. In addition to working to advance the rights of people of African descent, the movement also advocated the slogan "Black is Beautiful." Counter to prevailing ideas

FIGURE 15.1 Angela Davis
© AP Photo/Charles Bennett/PA Images

that black skin, features, and hair were ugly, the Black Power activists asserted their inherent beauty and encouraged black men and women to go "natural." The Afro was the predominant hairstyle in this movement, and several images from the 1960s feature activists such as Angela Davis with the distinctive look.

Here, *not* intervening in hair's natural growth was the way to make a powerful statement.

Feminist theorist bell hooks (1998) writes beautifully about the experience of managing black hair:

> Since the world we lived in was racially segregated, it was easy to overlook the relationship between white supremacy and our obsession with hair. Even though black women with straight hair were perceived to be more beautiful than those with thick, frizzy hair, it was not overtly related to a notion that white women were a more appealing female group or that their straight hair set a beauty standard black women were struggling to live out. While this was probably the ideological framework from which the process of straightening black women's hair emerged, it was expanded so that it became a real space of black woman bonding through ritualized, shared experience. The beauty parlor was a space of consciousness raising, a space where black women shared life stories—hardship, trials, gossip; a place where one could be comforted and one's spirit renewed. It was for some women

> a place of rest where one did not need to meet the demands of children or men. It was the one hour some folk would spend "off their feet," a soothing, restful time of meditation and silence. These positive empowering implications of the ritual of hair pressing mediate, but do not change negative implications. They exist alongside all that is negative.

Despite the history of the Black Power movement, many women—women of color and white women—chemically straighten their hair to achieve a sleek sheet of pin-straight hair deemed socially desirable. As hooks explains, while the homosocial space of a salon or barbershop can be a site of community building and social connection, beauty norms do not reflect the diversity of human shapes, forms, and colors. There are also bodily risks associated with the use of straightening agents on human hair—just as there are with the use of curling or "perming" agents.

Sociologist Barbara Katz Rothman's (2005) auto-ethnography of managing her adopted African-American daughter's hair is significant in its exploration of the intersections of gender, race, mothering, and social class. Rothman's personal and scholarly skills bring into focus many of the experiences of transracial adoptive families in the United States. Beginning with attempting to become adept at styling and caring for her 15-year old daughter's hair, Rothman writes about how lack of knowledge about hair is both materially and symbolically part of a racist history of American segregation. She adds,

> Hair is a test to be passed or failed, a trial to be endured, an intimate moment to be shared. In memoirs of those raised within the African-American community and those raised by white people, hair and the doing of hair emerges as a focal point for the discussion and for the experience of race.

Chemical politics of hair

For many of us, personal grooming and hygiene is indicated by how we style and care for our hair. Individuals can make or purchase countless styling products to clean, condition, defrizz, color, tease, calm, grow, eliminate, or relax hair. For example, lemon juice, coconut oil, olive oil, fruit, and beer are just some of the home-based "treatments" used to tame tresses.

Hairdressers also encourage customers to "use product" in order to achieve stylistic effects. The investment in "product" use has been explored by legal scholar Bridget Crawford (2011). Her research demonstrates that during economic downturns, women develop different strategies to maintain their hairstyles. Significantly, while white women develop strategies to save money by stretching out salon visits or foregoing hairstyling, African American women's consumption of hair care is less susceptible to economic cycles. Crawford argues this is an example of racial privilege for white women, whereby African American women must fit within a more rigid regimen of the presentation of self.

But what might be lurking in these conditioners, mousses, gels, and potions?

The Environmental Working Group, an organization that informs the public about health risks of pesticides and harmful chemicals and advocates for more sustainable production practices, has found that several popular brands of hair care products as well as top salons exceed the safety limits for use of formaldehyde. A carcinogen, formaldehyde may cause cancer or trigger asthma attacks, and the risk of illness increases over repeated exposures. This is clearly a risk both for the individual who is straightening her (or his) hair and for the stylist (often a woman) who works with these chemicals repeatedly. In an eerie twist, while chemicals used to treat hair can lead to cancer, cancer treatment can lead to hair loss.

Cancer treatment and hair

One of the most common treatments for cancer is the combinatory use of highly toxic drugs, or *chemotherapy*. These chemical treatments often have significant side effects that render the individual quite sick and physically incapacitated.

Alopecia, the loss of hair from the entire body including eyebrows and eyelashes, is a common side effect of drug treatments. Social psychological research has demonstrated how this hair loss is often devastating to the patient's self-esteem as well as troubling because their baldness may reveal their health status (Rosman 2004). These individuals are multiply stigmatized by being both bald and seriously ill. Even though wigs and headscarves can be worn to conceal baldness, often these "fixes" bring unwanted attention. In many cultures preoccupied by youth, health, beauty, and fitness, hair loss marks individual as sick, "outing" an individual's medical condition and potentially changing their relative status in social interactions.

Particularly in women, alopecia (medically induced or naturally occurring) can lead to emotional stress, reduced self-esteem, fear, and anxiety that can alter one's psychosocial state and affect social interactions and daily activities. Baldness and hair thinning are presented to women as if they are wholly unnatural, and yet less than 45 percent of women will go through life with a full head of hair (Dinh and Sinclair 2007). This prevalence actually means that it is *more* likely for women, as they age, to experience light to moderate hair loss.

However, many women and men who experience this form of hair loss elect to wear headscarves, hats, and/or wigs to repel attention. Human behaviors surrounding hair often conceal what is "naturally" occurring in favor of constructing a new natural, a thick head of hair, to manufacture and maintain.

Baldness as solidarity

Because hair loss is such a frequent side effect of chemotherapy and some other medical treatments, it is not uncommon for patients to completely shave their heads, so as not to have wisps and fragments, and for patients' loved ones and supporters to also shave in solidarity. Many women with breast cancer, for example, are supported by partners and family members who shave their hair off, too. Parents of children with leukemia will often shave their own heads to lessen their children's emotional suffering.

Here we see baldness not as a stigma, but rather a shaved head as a sign of love, community, support, and possibly hope.

Aging and hair

The World Health Organization reports that globally, life expectancy at birth is 68 years and ranges between 57 and 80 years depending on the income level of the country. The study of aging is termed gerontology. This field encompasses biology, psychology, medicine, and sociology, and examines how and why people change with age. Age is also a significant basis for social ranking, termed *age stratification*.

In some societies, the elderly enjoy high prestige. For example, in some pastoral societies, the elderly are called the *gerontocracy*. In most modern societies, however, the elderly are not well respected, with one of the few exceptions being Japan, where the elderly are considered wise and cultural values dictate respect.

The aging process affects hair in many ways. As we age, our hair color changes because we produce less melanin, a natural pigment that gives color to hair, skin, and the iris of the eye. Graying or whitening of hair (both scalp hair and pubic hair) is one of the significant indications of aging.

The transitions of aging are not limited to biological changes to one's hair or skin, but also include socially experienced and publicly documented transitions in status and social and cultural engagement. Gray hair, for example, might visibly symbolize one's change in social status. All elderly people may also face *ageism*, which is a form of prejudice and discrimination against the elderly.

For example, several people have studied the aging process as it relates to changes in social isolation and engagement (Victor, Scrambler, and Bond 2008), vulnerability to poverty and, finally, the intricacies of delivering and managing caregiving (Diamond 1995). The cultural practice of obscuring signs of aging evident in our hair is at least partially a reaction to the deep stigma associated with growing older.

Reversing the effects of aging?

As a means of mitigating ageism in the youth-oriented cultures of many countries, people engage in practices to appear younger, such as dyeing their hair or mitigating baldness. Indeed, 51 percent of people (of both genders) who dye their hair report doing so to cover gray hair, and they self-report that they believe coloring their hair enhances their romantic and professional prospects (Sherrow 2006). While men who fashion elaborate comb-overs to conceal a balding scalp may frequently be mocked, balding is nonetheless a very common phenomenon. The prevalence of male pattern baldness in men ages 40 to 49 is 53 percent, which means it is a more prevalent condition than nonbaldness (Rhodes and others 1998).

As we have explored in other chapters, *medicalization* is a process whereby a condition or behavior is categorized as a medical problem or a disease to be treated. As we addressed above, medically-induced alopecia is specifically introduced by certain interventions, but there are far more individuals (overwhelmingly men) who experience baldness as a normal part of the aging process. Some pharmaceutical options, such as over-the-counter Rogaine® and the prescription drug Propecia, are used to "treat" male pattern baldness. These treatments have adverse side effects that range from skin irritation to the risk of impotence, abnormal ejaculation, and prostate cancer.

Men, and sometimes women, may also elect to have hair implants, sometimes called plugs, to surgically "reverse" their baldness. This procedure is available only to those people who still evidence at least some healthy and thicker regions of hair on their head. Follicles are harvested from a healthy area of the patient's head and then, using small needles, these grafts are punctured into new locations on the head. It is difficult to ascertain how prevalent hair transplantation is, but it is very expensive ranging between US$5000–12,000, depending on the procedure and the amount of hair.

So important culturally and sexually is the presence of hair on our heads, that scientists have developed a way to implant follicles and grow human hair on the backs of lab mice. Big Pharma, as well as those suffering from the stigma of balding, have much to gain from these cross-species scientific endeavors. However, in the absence of a feasible way to transplant mouse follicles onto human heads, as the next section shows there is a thriving market in human hair that spans the globe.

The hair trade

Hair extensions are lengths of real or artificial hair that are attached close to the scalp through wax or glue, or through a weaving technique. The desire for long hair, along with fashion and entertainment industries driving consumption for instant gratification in hairstyles, has created an increased demand for human hair. Just as there is a robust trade in human tissues and organs, there is also considerable traffic in human hair.

The global hair trade, considered a rapidly growing market, is a transnational site of commerce where typically poor women sell their hair to brokers who deal in pounds (weight) of hair. India is one of the top exporters of human hair. The United States, China, and the United Kingdom are the top importers.

Journalist Scott Carney, author of *The Red Market*, investigated a temple in India that provides "remy" (or "high grade" all-human) hair to Western markets, typically elite markets for use in photo shoots, among models and celebrities, and the wealthy. Members of the temple are not informed that their hair, which they've shorn for religious purposes, is being sold for trade, with a profit of US$10-12 million per year.

Cultural critic Esther Berry (2008) has investigated the world's largest multinational natural hair company, Great Lengths, headquartered in Rome, Italy. She writes (at 63):

> [T]he global exchange of human hair, a business implicated in structuring First and Third World femininities, is based on consumer and producer relations between the West and the rest, as hair is culled and then spun into a "repulsive gift" *hair extensions* bestowed upon the scalps of a First World clientele. Indeed, in many ways, consuming the hair of the Other contributes to reinscribing colonial boundaries, as the First World's hair-extended women play out a narrative reminiscent of Rapunzel in which they clamber towards alterity. Up the West's ivory tower, this is a climb mounted on borrowed braids.

Berry suggests that the hair market is one in which privileged women purchase their beauty accessories, in this case human hair, from other women who will not be able to then participate in the beauty standards under which they are similarly judged. As widely reported, many women (and men) sell their hair under conditions of economic deprivation and scarcity (Khaleeli 2012).

FIGURE 15.2 Human hair trade
Image courtesy of Adrian Fisk
At a hair processing factory in Eluru in the state of Andhra Pradesh, a new consignment of village hair has arrived and is about to be sorted. Often children are used to untangle the tightly knotted village hair as their fingers are small and nimble. http://www.flickr.com/photos/adrianfisk/3403976967/

Examining the hair trade, which at heart is about artifice, reveals some of the ironies of trying to appear racially or sexually authentic, or a "natural-born" beauty. The use of hair extensions is often shrouded in secrecy, with women reluctant to admit their use of another's hair. Essayist Lisa Jones writes: "If hair is the key racial signifier after skin, then the trade makes a fine mockery of it. Processed Asian hair passes as black hair. Italian stock is allegedly blended with hair from the Third World and this passes as European . . . most of what the hair trade sells is not dead straight hair—it's "texturized" hair; hair that is quite particularly African American/African-diaspora hair, hair that is emulated in perms by European women worldwide" (Jones 1997: 296).

In many respects, the transnational hair trade seems a very long way from the Black Power movement of the 1960s and the primacy of the "natural" Afro. The intersections of race, class, gender, and geography lead to complex negotiations of "beauty" from women (and sometimes men) who are differently situated in hierarchical power structures and transnational economies.

A CLOSER LOOK

"Manscaping"—or, changing fashions in men's body hair

For many women, particularly younger generations, managing body hair has become a frequent, often tiresome routine of tweezing, plucking, shaving, trimming, and waxing. Until recently, most men have been free from these routines—except for managing their "five o'clock shadows" through facial shaving or waxing. While it is true that historically Western men may have groomed their body hair regularly, it has not been the case across the past century.

The rise in male grooming can be attributed to the contemporary phenomenon of *metrosexuality*. A metrosexual is a man who is concerned with grooming, appearance, fashion, and aesthetics, who is typically not identified as homosexual. Although some homosexuals also identify as metrosexual, what is interesting about metrosexuality is the way it "blur[s] the visual style of straight and gay in a restless search 'to spend, shop and deep-condition'" (Miller 2005: 112).

The increase in hair removal from the chest, back, buttocks, and genitals of men is often attributed to the availability of and rise in web-based, easy-to-access pornography that depicts typically hairless men (Ramsey et al. 2009). Several men self-report that their genitals appear larger (to them) when they shave or trim their pubic hair.

Matthew Immergut (2010) has examined the interrelationship between a form of metrosexuality, *manscaping*, that is the waxing, shaving, lasering, or simply shaping any hairy region of the male body below the head, and the ecological practice of landscaping. Immergut analyzes how gaining control over the natural (be it bodies or terrains) is part of becoming "civilized" as a person or an ecological landscape.

So as men in industrialized countries have become more "civilized" and move out of blue-collar professions with global marketplace changes, they are also becoming more manicured, like suburban lawns. Their previously "unruly" body hair is brought under control through social and technical management, just as the land is tilled to make it usable for modern agriculture.

Conclusion

To summarize, human hair is much more than ingrown hairs, split ends, graying temples, and comb-overs. It is, in part, big business and global trade organized around pharmaceutical, beauty, and commercial interests. While we have a personal relationship with our own sense of style and the grooming habits we practice, these styles and practices are not of our own making. They are part of larger historical and cultural trends including ideas about gender and race, norms of sexuality, and shifting beauty ideals.

Like other parts of the body, hair can be modified: shortened, lengthened, colored, concealed, textured, styled, or completely removed. How we present ourselves through our hair says a great deal about how we conform to social mores, or not. Long blond tresses can signal luxurious beauty and (white) femininity as per Western dominant ideals, just as hot-pink or chartreuse cropped hair can signal our disavowal of those very ideals. The "beauty" of hair is that it is so malleable, unlike some other parts of our body that are more resistant to change.

See also

Organs; Faces; Skin; Genitals

Key terms

Sexually dimorphic, chemotherapy, alopecia, hair trade, man-scaping, age stratification, gerontocracy, ageism, medicalization, metrosexuality

Discussion questions

1. Consider your own practices and routines regarding your hair: scalp, facial, body, and pubic. How much time do you invest in your hair every day, week, month? How much money do you invest? For whom and why do you think you are managing your hair?
2. Considering the global hair trade, what are some potentially troublesome issues invoked by the buying and selling of human hair? What are some of the ironies? Is there a way, as some have suggested, to buy human hair ethically? What might this look like?

3 Social expectations about human hair vary depending upon a person's age, race, gender, sexual orientation, religion, geography, and social class. What are the current hair trends for different demographic groups in your country or region? Discuss what these hairstyles signify and how an individual achieves this "look."
4 In addition to stacked-abs exercises, teeth whitening kits, and penile enhancement pharmaceuticals, manscaping is part of the ever-expanding list of tasks in the male body project. What are some of the components of today's manscaping projects? How might this vary from culture to culture?
5 What are some ways hair is managed as we age? Who benefits from this management of graying and thinning hair? What are the risks to "fixing" the problems of baldness and grayness? What do you anticipate will be the next trend in hair as medicine and business collaborate to find new solutions and markets?

chapter 16

Breasts

Introduction

Mammaries. Lady pillows. Boobs. Milk sacks. Hooters. Ta-tas. Airbags. Knockers. Headlights. Bumpers. Tits. Busters. The twins. Yabbos. Balcony. Grab bags. Jugs. Knobs. Chesticles. Pechos. Bazooms. Melons. Pair. The girls. Rack. Fun bags. Chi-chis. Breasticles. Peaches. Mangoes. Chestnuts. Breasties. The ladies. Goosebumps. Nunga-nungas. May-mays.

These are just some of the gendered nicknames for women's breasts. While a few of these monikers are related to function (e.g., "mammaries," "milk sacks"), most speak to the central place of women's breasts in sexuality, often as defined and labeled by men and related to male desire.

In 1758, the Swedish scientist Carl Linnaeus described a class of animals as "*mammal*," from the Latin *mammalis*, or "of the breast," and *mamma*, "breast." In one subjective classificatory move and with the weight of scientific authority behind him (Schiebinger 1993), he linked breasts to sexuality to motherhood to human evolution.

Linnaean taxonomy was hugely influential, designating living organisms into "kingdoms" of animal, vegetable, and mineral each with its own internal classifications. What we have come to know as "human" in relation to "animal" and "mineral" was deeply shaped by Linnaeus (Haraway 1988). Ideas about gender, sexuality, and race are woven into these scientific understandings.

Physiologically, breasts are organs located atop the pectoral (or chest) muscles and are made up of connective tissue, fat, lobes, lobules, ducts, and lymph nodes. Male breasts are similar to female breasts, but without specialized lobules for milk production.

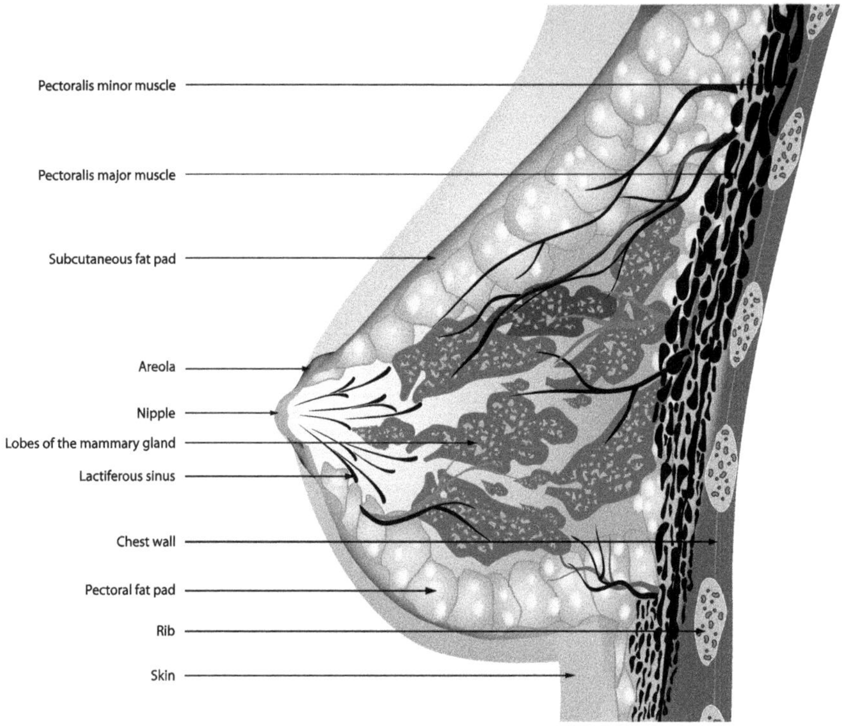

FIGURE 16.1 Anatomy of the breast
© Shutterstock

The lobules branch out from the nipples like a river delta, and each lobule contains tiny sacs called alveoli. Lobules are linked by a network of ducts, which in lactating women carry milk from the alveoli to the areola (center portion) of the nipple. Fat, ligaments, and connective tissue fill the spaces around the lobules and ducts.

Breasts vary in size, shape, and appearance, both in relation to other breasts and across the life course. Humans, who most typically birth single babies, almost always have two breasts, while other species often have more, consistent with birthing larger numbers of offspring with each pregnancy (e.g., dogs can have up to 12 teats). Girls menstruate and develop breasts at different ages, and throughout their lives women's (and men's) breasts may be affected by health status, reproductive experiences, sporting activities, surgical history, and many other factors.

What remains fairly constant in many cultures is the potent connection between breasts, femininity, sexuality, and body image. Perhaps more than any other organ—even genitals—a woman's sexual sense of self may be tied to her breasts, a link fostered through social practices, cultural representations, and gendered body politics.

Lactation politics

Given the "functional" nature of breasts as a source of food for infants, you might think that breastfeeding is apolitical or non-controversial. But you would be wrong. Breastfeeding has emerged in the last century in the Global North as a highly significant social issue, in part because the practice embodies changing and contested meanings of sexuality, gender, motherhood, and modernity. Sociologist Linda M. Blum writes (2000: 2), "Conversations that look 'at the breast'. . .are not private. Motherhood and breastfeeding have been and continue to be public matters."

Of course, breastfeeding practices have long reflected social inequities. Historian Janet Golden (2001) traces the use of "wet nurses," or lactating (milk-producing) women who breastfed other women's infants, from colonial America through the 20th century. Reasons for the use of wet nurses varied, but they most often reflected class differences, with women of higher economic status utilizing the embodied labor of women of lower economic status. In imperial contexts, indigenous women were often used as wet nurses by white colonials (Chaudhuri 1988). In the U.S. slave economy, African American women were forced to nurse the infants of white slaveholders, often at great cost to relationships with their own children (Jones-Rogers 2011).

In contemporary industrialized societies, breastfeeding has been a component of the "mommy wars" that have erupted between women who work and those who stay at home raising their children (Steiner 2007). Debates have been vociferous, with working moms vilified for subjecting their children to paid child care and formula, and stay-at-home moms criticized for putting their children's needs before their own. Working moms are criticized by some feminists for interrupting breastfeeding or failing to try nursing in the first place.

Indeed, when Lisa Jean (Moore 2007) was teaching an undergraduate sociology class, her leaking breast milk led to commentary from several students about the appropriateness of a nursing mother being a professor. These varied beliefs about maternalism, breasts, and differences among women, as well as cultural confusion about women's place in the workforce, demonstrate cultural ambivalence about female power.

However, as Susan Douglas and Meredith Michaels argue (2005), the debates about women's roles ultimately undermine all women by idealizing motherhood. The debates also often fail to situate the "mommy wars" historically and in the context of race and ethnicity. Whereas some women, especially middle- to upper-class women, have the luxury to stay home and criticize women who work, women in the working and service classes do not. For several decades, even achieving a lower middle-class standard of living in industrialized economies has required two working adults. Thus, many women—single mothers, poor women, African American women, undocumented immigrants, indigenous women—are left out of the "mommy wars" entirely.

One site of concern about breastfeeding is who does it and for how long, or the rates. Although scientific evidence seems to favor breastfeeding over formula for child health and survival (e.g., Wright and Schanler 2001), not all women breastfeed. The World Health Organization recommends that women breastfeed exclusively for six months and that they try to continue nursing their children for up to two years. Breast milk contains antibodies and other substances to ensure infant and child health—ingredients missing from manufactured infant formula.

The WHO Global Data Bank on Breastfeeding shows that rates vary dramatically from country to country and region to region, with some women exclusively breastfeeding for many months, other women nursing their children for years, and some not breastfeeding at all. Breastfeeding rates are increasing in several countries, in large part—according to WHO—owing to campaigns about breastfeeding's benefits and institutional (e.g., hospital) changes including additional support for new mothers. International NGO La Leche League works to increase rates of nursing through outreach, support for women, and education.

Not everyone thinks breast is best, however. Scholar Joan Wolf (2010) takes issue with campaigns for breastfeeding, suggesting that not only is the data inconclusive about the benefits, but also that imperatives to breastfeed may harm women who either cannot or choose not to do it. In addition, concerns about contaminated breast milk may lead many women to avoid nursing their infants. Breast milk has been found to be full of toxins, and is now an ideal substance for the practice of biomonitoring, or measurement of pollutants in the body (Casper and Moore 2009). Some feminist critics also claim that breastfeeding keeps women tied to their bodies and reproductive roles, echoing earlier essentialist arguments.

Whether one comes down on the side of breastfeeding or formula, one thing is certain: framing the issue as one of individual choice—as many debates in the Global North do—precludes consideration of the *structural* aspects of nursing, including women's roles in society, familial and institutional support for breastfeeding, cultural norms, economic issues, women's health and wellbeing, geopolitics, and global poverty. One might ask, instead, are women located in geopolitical and economic contexts that advance their health and that of their offspring, whether or not they "choose" to breastfeed?

A CLOSER LOOK

Picturing breasts

Breasts are visible even when they are not; that is, although they are typically covered up in many (but not all) cultures, they often appear as "curves" or "mounds" of presumed womanhood beneath fabric. They are seen as a representation of sexuality—and many women maximize their visibility and appeal through artfully arranged necklines, push-up bras, cleavage, and other techniques. Alternately, in some cultures, women minimize their visibility through loose clothing.

However, there are other ways to "picture" breasts, too, including *mammography* and art.

Mammography is a screening and diagnostic technology that relies on x-rays to examine breasts, particularly for signs of precancer or cancer. Breast cancer is the leading cancer among women globally, with rates increasing in the Global South owing to increased urbanization, adoption of Western lifestyles, and insufficient resources to manage care (Anderson et al. 2011).

During a mammogram, with the help of a technician a woman places her breast on a plate connected to a machine. Another plate is lowered to compress the breast tissue to a flat layer. Several images are taken throughout the procedure, requiring the woman to manipulate her chest, arms, shoulders, and breasts in relation to the machine. Women who undergo mammography often describe the experience as uncomfortable, humiliating, and even painful (Davey 2006).

In the U.S., Canada, and Europe, women 50 years of age and older are recommended to have mammograms every two to three years. In many poorer regions in the world, mammography is not widely available, it remains expensive, and guidelines are quite varied (Anderson and Jakesz 2008). As clinician Peggy Porter (2008) notes:

> "current global figures cannot truly reflect the underlying economic and cultural diversity driving increased incidence and related mortality . . . The challenge is knowing whether the successful strategies refined in predominantly white, affluent populations (early detection through mammography, targeted hormonal and anti HER2 therapies, and improvements in chemotherapy) will be effective, or even possible, in lower-income countries with different racial, ethnic, and cultural make-ups."

For centuries, artists have also imaged women's breasts, portraying them in terms of sexuality, desirability, violence, beauty, aging, and maternity. Contemporary Brooklyn-based artist Clarity Haynes has created the Breast Portrait Project, a series of drawings of women's torsos including their breasts.

In her artist's statement, Haynes writes: "Historically, the painted portrait signifies privilege, beauty, and social power. I choose to portray real women's bodies, which are often invisible in this culture . . . The face is our commonly recognized self—our 'mask' of identity. Focusing exclusively on the torso shines light on a personal and private part of the individual that is usually hidden."

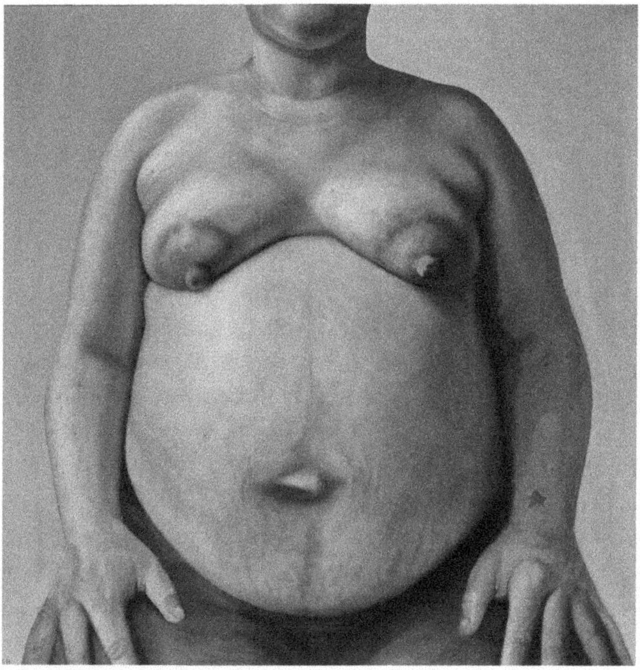

FIGURE 16.2 Torso portraiture
"Lyz", oil on linen, 58" x 58", Clarity Haynes

Images of breasts, whether radiographic or artistic, are invariably seen through the particular cultural lens of the viewer. Even the "objective" image of a mammogram is interpreted through and in relation to a set of meanings about gender, bodies, and health. Images can bolster existing beliefs about and norms surrounding women's bodies and visualization, or they can—like Clarity Haynes's work—explode these norms.

Breasts, gender, and sexuality

The relationship between breasts and cultural definitions of femininity or womanhood is persistent in many cultures. While the 18th century saw the rise of scientific classification, our contemporary moment—especially in the Global North—is characterized by a visual proliferation of breasts, from magazine covers to fashion to Internet porn.

The 1940s saw the emergence of the term "pin-up girl," denoting a particular visual representation of women's bodies emphasizing hips and breasts. Sometimes nude, often scantily clad, the "pin-ups" were designed for display and literally pinned to a wall in a teenage boy's (or girl's) bedroom or a workplace. Pilots in World War II often detailed their airplanes with similar images of women, suggesting provocative associations between militarization, masculinity, and cultural representations of female sexuality.

Pornography is the visual display or explicit description of sexual activity and bodies, specifically breasts and genitals. Such representations are as old as humans, but modernity brought about new ways to display and share images. Print made possible the widespread distribution of magazines, posters, pictures, and books, while the Internet has led to a proliferation of sites offering every manner of sexual display. In *Woman*, journalist Natalie Angier (1999: 140) writes: "What we love is not the breast per se but the fantasy breast, the aesthetic breast of no practical value."

Given that pornography displays predominantly women's bodies, especially breasts and genitals, critical attention has emphasized its potential harm to women including objectification and inducement to violence against women (Dworkin 1981). Pornography became a central theme in the feminist "sex wars" of the late 1970s and 1980s, as some feminists argued for women's pleasure and agency in pornography and sexual activity (Duggan and Hunter 1995). These feminists, while diverse in perspective and identity, are typically understood as "pro sex" feminists.

Particularly interesting are the ways that non-pornographic materials have been read and consumed for sexual pleasure. For example, generations of girls and boys especially in the Global North have looked to *National Geographic* for pictures of breasts, penises, and naked bodies. Anthropologist Catherine Lutz and sociologist Jane Collins (1993) offer a critical analysis of the magazine including its racial and gendered dimensions, while other critics have termed *National Geographic's* images of non-white ethnic others "tribal nudity."

Often missing from conversations about the cultural politics of breasts and male objectification via pornography are women's own subjective experiences. This includes women's sexual, aesthetic pleasure as it relates to their breasts. While there are shelves and shelves of books about breast cancer and breast health, there are almost no books that explore breast pleasure *for women*.

Sexual pleasure is especially "taboo" when women are nursing their babies—despite ample evidence that breastfeeding can be deeply pleasurable. Philosopher Iris Marion Young (1992: 160) writes: "Breasts are a scandal because they shatter the border between motherhood and sexuality. Nipples are taboo because they are quite literally, physically and functionally, *undecidable* in the split between motherhood and sexuality."

Breasts, beauty, and body image

Given their ubiquity in popular culture, breasts are often a source of anxiety for girls and women who, in their everyday lives, must contend with the emergence of highly sexualized body parts. We might wonder, as girls, when our breasts will develop: Will they be too small? Too big? Why have some of my friends already gone through puberty, yet I haven't? Now that I have breasts, why are all the other kids either staring at me or avoiding me? Can I still do sports? What will I wear? How do I know what size bra to buy?

Men, too, may worry about the size and shape of their chests. The term "man boobs" is used disparagingly to describe breasts on men, especially men who are heavier than is considered normative. Whereas in some cultures, a smooth, hard chest is designated as the ideal male physique, any protuberance that bounces and might be considered a "breast" is suspect.

At the heart of this cultural frame is the gender binary, and the surveillance and policing that goes into ensuring that "girls" and "boys" and "women" and "men" are properly performing their gender. The Internet is full of sites, including men's "health" magazines, that offer tips on how to rid these "undesirable" features from a man's body. Clearly, to be manly in certain cultures is *not* to have breasts.

For many women—and men—who wish to change the size and/or shape of their breasts, or who are recovering from breast cancer treatment, surgical options abound, at least in resource-rich regions of the world. Increasing breast size is called *breast augmentation*, while decreasing breast size is called *breast reduction*. Reasons for surgery vary, although typically they relate to cultural ideas about body image, self-esteem, sexuality, and femininity (Wolf 2002 [1991]). Obviously, in resource-poor regions of the world, cosmetic breast surgery is out of reach for all but the wealthy, and even basic breast cancer management (i.e., early detection and treatment) is challenging (Anderson et al. 2011).

Breast augmentation is the most common form of elective plastic surgery in many countries, although liposuction/lipoplasty (the liquefaction or removal of fat beneath the skin) has edged out breast augmentation as the most statistically popular procedure worldwide (ISAPS 2012). The nations with the highest rates of plastic surgery in 2012 were, in order, South Korea, Greece, Italy, Brazil, Colombia, the U.S., and Taiwan. More than 300,000 women in the U.S. underwent breast augmentation in 2012 (Youn 2013), and almost a third of these involved reconstructive surgery after breast removal owing to cancer. Feminist critique has centered on the patriarchal imperative that women succumb to breast reconstruction (Lorde 1980).

Women also may opt to *reduce* their breast size, typically because they have very large breasts that lead to discomfort and back pain or because their breasts impede active participation in sports. The latter motivation invokes the legend of the Amazons, a race of "warrior women" in classical antiquity. The word amazon is derived from *a-mazos*, literally without breasts. The Amazons were believed to have their right breasts cut or burnt off so they could more effectively throw javelins (Davis-Kimball and Behan 2003).

Breast surgery is a key component in possible options for transgender body modifications, sometimes called *sex reassignment surgeries* or *gender confirming surgeries*. "Top surgery" is the procedure by which trans men undergo mastectomy and surgical sculpting of the chest, typically in conjunction with hormone therapy (Currah 2008). Trans women often undergo breast augmentation, also typically in conjunction with hormone therapy. In both cases, breasts and chests are altered to align with the chosen gender identity. Here, as with the "man boobs" example above, breasts signal femininity/womanhood, while lack of breasts signals masculinity/manhood.

There are numerous feminist critiques of elective plastic surgery, many of which focus on the ways in which women (and increasingly men) are culturally coerced into various body modifications in the name of beauty (Heyes and Jones 2009). Researchers have found that people who have one procedure often go on to have additional procedures, turning them into "surgery junkies," in sociologist Victoria Pitts-Taylor's words. She goes on to suggest, however, that feminist critiques often portray women as "dupes" of the beauty industry, or as suffering from "bodily self-hatred as an effect of the pressures of patriarchal culture" (2007: 20). The story, she suggests, is more complex in that women have varied relationships to interventions and enjoy more agency than some feminist critiques credit.

While the sociological story of cosmetic surgery may be more complicated than many scholars portray it, the health consequences of breast implants are a serious issue. In the early 1990s, controversy erupted around silicone breast implants, in particular their safety (Jacobson 1999). The U.S Food and Drug Administration (FDA) curtailed the use of implants outside of controlled clinical trials (Kessler 1992), evoking heated debate among companies such as Dow Corning (which ultimately went bankrupt as a result of lawsuits), government officials, and women's health advocates.

Although there are new kinds of breast implants on the market, health concerns remain. Pain, rupture, and the need for additional surgery often follow breast augmentation. There is some evidence that breast implants are associated with autoimmune disorders, although overall, despite the popularity of breast implants, there is inadequate safety data. The transnational women's health NGO Our Bodies, Ourselves maintains a page on body image and breast implants at its website: http://www.ourbodiesourselves.org/book/companion.asp?id=1&compID=30.

An important feminist concern is the cost of rigid beauty standards and bodily interventions related to the breast, especially for young women, as teenagers are a fast-growing segment of the breast surgery industry. For example, Italy banned plastic surgery for minors out of concern for the rapidly escalating numbers of girls seeking breast augmentation and other procedures.

"Indecent" exposure and public breasts

Ironically, given the connection between breasts and sexuality one might expect that breasts are everywhere. However, the flip side of sexualizing breasts is that they must, in many cultures, also be covered up or contained. Indeed, it is in most countries *illegal* for women to expose their breasts in public, as for example through openly nursing their babies or sunbathing topless.

Breast exposure by women in many settings is considered to be in violation of public decency and lewdness laws or religious doctrine, and also an incitement to sexual assault by men (who are perceived as being unable to control themselves when faced with naked breasts). Men, on the other hand, may go topless almost universally. In many indigenous cultures in warmer climates, it is normative for women's breasts to be exposed and often adorned.

The history of keeping breasts hidden or contained is also the history of undergarments, and specifically the bra and its precursors. For centuries, women used corsets made of fabrics, whale bone, wood, and other materials to contain their breasts. Corsets both covered and amplified the breasts while also cinching the waist tightly, thus creating a normatively "womanly" or curvy form. The early 20th century saw the rise of the brassiere. The bra, as it came to be called, was patented by Mary Jane Phelps, a New York socialite, and later sold to the Warner Brothers Corset Company.

Historian Jill Fields, author of *An Intimate Affair: Women, Lingerie, and Sexuality*, describes and analyzes the many iterations of lingerie, including bras, across the decades. She writes: "underwear, although worn next to the body and thus ostensibly hidden from outside view, is a crucial part of the gendered fashion system. Private and sexualized, yet essential to the shaping of the publicly viewed silhouette, intimate apparel—a term in use by 1921—is critical to making bodies feminine" (2007: 2–3).

Feminists and others, through various forms of activism including public nudity, have challenged public decency laws and the imperative to contain breasts. Criticism of "decency" laws has focused on the gendered politics of modesty, in which women are to be covered up to prevent men from looking at their bodies. Of course, the onus is on women to protect their bodies through not revealing themselves, lest they incite harm against them. The parallel gendered assumption is that men cannot control themselves. Such logic undergirds use of the hijab in Islam, although women's reasons for wearing the veil are complex (Ahmed 2011).

La Leche League International has worked across the globe to increase support for breastfeeding, including legally challenging decency laws that force women into unsanitary bathrooms and other "private" spaces to nurse their babies. The organization's website includes a page entitled "Breastfeeding and the Law" with sections on laws in the U.S. and outside the U.S. Mothers' activism on behalf of nursing their babies generates significant media attention; in 2013, a YouTube video featuring British poet Hollie McNish defending breastfeeding "went viral."

The Ukrainian feminist group FEMEN has made toplessness a key feature of its "sextremist" activism. Through a variety of demonstrations in several countries, FEMEN has used women's nudity, and particularly breasts, to challenge patriarchy by upending its rules about modesty and sexuality. On its Internet site, the group describes itself as "the scandal famous organization of topless women activists, who defend with their breast sexual and social equality in the world . . . FEMEN is the new Amazons."

Yet not everyone appreciates this brand of activism; the group Muslim Women Against FEMEN challenges the group as Islamophobic and imperialist; that is, as imposing Western views that position Muslim women as in need of saving.

Breast cancer

The term "breast cancer" strikes fear into the hearts of many women, especially in the developed world. In the Global South, while breast cancer

FIGURE 16.3 Femen activists
Image by Jacob Khirst

is a significant problem, other issues such as hunger and infectious diseases are often experienced as more urgent and more deserving of scant public dollars. Yet, the World Health Organization reports that breast cancer is the most prevalent cancer in the world. In 2008, approximately 1.4 million women around the world were diagnosed with breast cancer, and breast cancer was the most common cause of death in women.

Physiologically, breast cancer is a malignancy in the cells and tissues of the breast. Cancer cells, characterized by unchecked cell division and proliferation, are often clustered into a mass recognized as a tumor. However, not all tumors in the body are malignant or cancerous. Breast cancer may spread into lymph nodes and into other parts of the body; this is called *metastasis. Ductal carcinoma* begins in the ducts and invades surrounding tissues; about 80 percent of breast cancers are of this type. About 10 percent of breast cancers are *lobular carcinoma*, which begins in the lobules and invades surrounding tissues.

Breast cancer is diagnosed with mammography or breast ultrasound, and also by women themselves, who may notice breast changes such as lumps, or their healthcare providers. Treatment includes surgery (e.g., removal of cancerous tissue and sometimes the breast itself), chemotherapy, hormonal therapy, biological therapy, and radiation. Critics of the "war on cancer" note that these treatments are often very good for industry—with some of the same companies producing both industrial chemicals and chemotherapy—and do little actually to prevent cancer or reduce incidence (Davis 2009).

Like so many diseases, breast cancer is stratified, although not in the ways we might typically expect. The incidence of breast cancer is higher in Western Europe, North America, and Central and Eastern Europe—the so-called developed world—and lower in Africa and Asia. Yet *mortality*, or deaths caused by breast cancer, is higher in Africa and other low-resource regions. The higher incidence is owed, in part, to access to the screening technologies that locate and diagnose cancer (e.g., mammography), and also as a consequence of "lifestyle" (e.g., diet, alcohol, stress) and heredity factors (Althuis et al. 2005).

Higher mortality in the Global South has a great deal to do with lack of access to screening and early intervention technologies (Anderson and Jakesz 2008). Even in the Global North, poor women and those without access to care have higher *morbidity* and mortality rates. In the United States, researchers at the University of California found that poor, minority women without health insurance were the most likely to delay treatment and thus die from breast cancer (Mientka 2013). A 2013 study found persistent differences between black and white women in the U.S. beginning in the 1980s, with black women experiencing higher mortality rates due to disparities in access to screening and intervention (Mandelblatt, Sheppard, and Neugut 2013).

A robust movement of breast cancer activists emerged in the 1980s, in response to biomedical approaches, the cancer-industry complex, and lack of action on women's health care (Knopf-Newman 2004). Sociologist Maren Klawiter (2008) analyzed various dimensions of this movement including grassroots patient empowerment, environmental justice and breast cancer, and growing awareness of corporate-funded breast cancer work. Her research makes clear that breast cancer is among the most politicized women's health issue in the Global North, along with reproduction. Breast Cancer Action in San Francisco, California epitomizes a grassroots approach, one that has been deeply attentive to issues of access and race and also critical of biomedicalization.

In *Pink Ribbons, Inc.*, Canadian scholar Samantha King (2008) draws attention to commercialization of the breast cancer movement, and the ways that corporate sponsorship has led to emergence of "survivorship" as a social, but not necessarily feminist, imperative. The search for a "cure" for breast cancer, she argues, undermines prevention efforts, including examination of environmental causes of the disease. Sociologist Gayle Sulik (2012) offers an insightful look at the corporate interests behind the ubiquitous pink ribbon, arguing that philanthropic and commercial efforts to "think pink" detract from prevention efforts and women's health care needs, ultimately leading to higher morbidity and mortality rates.

Although breast cancer is widely considered a women's disease, it can also affect men and may be especially stigmatizing for them. Breast tissue in men is typically undeveloped due to androgenic hormones (e.g., testosterone) produced by the testicles. Men have fewer ducts than women and no lobules. While cancer cells may form in the ducts, the different structure of men's breasts leads to much lower rates of breast cancer. Male breast cancer is most common in older men, with most cases diagnosed in men between ages 60 and 70.

Globally, less than 1 percent of breast cancers occur in men. In the United States, the American Cancer Society estimates about 2200 new breast cancer cases diagnosed in men, with approximately 410 deaths from breast cancer. Because breast cancer is framed largely as a women's disease, men may be less likely to seek diagnosis and treatment (Iredale and others 2005).

It is also important to note that breast cancer affects transgender people. For transgender women, breast development and hormone therapy can increase cancer risk. For transgender men, high levels of testosterone may convert to estrogen in the body and thus lead to an increased risk of breast cancer. For both transgender men and transgender women, interactions with the health care system may be uncomfortable, leading to avoidance of diagnosis and treatment. In addition, LGBT people are often underrepresented in public health and clinical research (Boehmer 2002).

A CLOSER LOOK

Angelina Jolie's mastectomy

On May 14, 2013, international celebrity Angelina Jolie ignited a firestorm of media coverage by penning an op-ed in the *New York Times*, in which she revealed that she had undergone a double *mastectomy* to prevent breast cancer. Results of BRCA1 testing indicated a high risk for Jolie of developing breast cancer, a disease that had already claimed her mother. She opted for the most aggressive form of prevention.

BRCA1 and BRCA2 are genes that produce proteins capable of suppressing tumors. Mutations in these genes can interrupt DNA repair and lead to the development of cancerous cells. BRCA1 and BRCA2 are inherited, and together, mutations in these genes account for up to 25 percent of hereditary breast cancer and 5–10 percent of all breast cancer. Mutations in BRCA1 and BRCA2 also account for up to 15 percent of ovarian cancer, according to the U.S. National Cancer Institute.

Availability of testing for mutations in BRCA1 and BRCA2 has led to a proliferation of prophylactic (or preventive) interventions, raising ethical questions such as the costs and benefits of aggressive treatment in the absence of disease (Surbone 2011). The test also provokes legal issues: in 2013, the U.S. Supreme Court ruled that isolated human genes, such as BRCA1, cannot be patented while simultaneously ruling that any *manipulation* of the gene can be patented.

In her *New York Times* opinion piece, Jolie thanked her partner, actor Brad Pitt, for his love and support, and also noted that the decision made her feel "empowered." She offered a timeline of the various procedures involved, including the insertion of implants, noting that "the results can be beautiful."

Jolie's aim in writing the article was her belief "that other women can benefit." She wrote: "For any woman reading this, I hope it helps to know you have options." Of course, Jolie had options not available to many women even in the United States, much less around the world. As a wealthy, white, heterosexual woman in a nation with extensive health care resources for those who can afford them, Jolie was able to secure BRCA1 testing and, presumably, the best breast surgical and follow-up care money can buy.

Jolie's critics noted some of these health care disparities—breast cancer is the leading cause of cancer deaths in women globally and basic early detection technologies are not widely available—as well as gendered aspects of emphasizing body image and beauty through breast reconstruction. Some worried that Jolie's disclosure would encourage testing and unnecessary radical surgery.

Jolie's was not the first high-profile celebrity announcement of breast cancer. In 1974, the First Lady of the United States, Betty Ford, underwent a radical mastectomy, fostering much public discussion. Communications scholar Tasha Dubriwny (2009) argues that Ford's surgery and its coverage in "minute detail" led to emergence of the figure of the breast cancer "survivor"—a figure that remains quite robust today. Ford was universally portrayed in newspapers as an ideal and compliant (i.e., feminine) patient.

Thus, while both Betty Ford and Angelina Jolie helped to raise "awareness" of breast cancer in the United States and indeed globally given their mass popularity, as Dubriwny notes awareness is insufficient. Public and media awareness alone—of breast cancer or any other disease (Lerner 2009)—does not mitigate social inequality or the global systemic problems that lead to stratified morbidity and mortality.

FIGURE 16.4 Angelina Jolie
© PA/Press Association Images
Celebrity Angelina Jolie in her first public appearance after undergoing a double mastectomy.

Conclusion

Breasts serve an important physiological function in the reproduction and maintenance of human beings, through lactation and breastfeeding. So important is the breast that Linnaeus crafted the word "mammal" on the basis of this function.

Yet as we have seen in this chapter, breasts are also highly social body parts—garnering desire and disgust, shame and pride. Breastfeeding, representations of breasts, and breast cancer diagnosis and treatment are rife with cultural meanings, and each of these activities is politicized in ways that reflect social understandings of gender, femininity, and sexuality.

Human beings have gone to great lengths to control representations and uses of their breasts, while at the same time, images of breasts and beliefs about them end up transfixing many. Many cultures appear to be breast obsessed in ways that reveal gender dynamics. We would do well to attend to the construction of rituals, laws, medical practices, entertainment, and art that fixate on these fleshy body parts.

See also

Cells and tissues; Organs; Fluids; Endocrine system and metabolism; Digestion; Reproduction; Genitals

Key terms

Mammal, mammography, pornography, breast augmentation, breast reduction, sex reassignment surgery, gender-confirming surgery, mortality, morbidity, metastasis, mastectomy

Discussion questions

1 When did you first become aware of breasts, your own or others? What did you think about them? Why do you think you had those thoughts?
2 Breastfeeding, topless sunbathing, and breast pride activism are all reasons why women's breasts might be bared in public. How do you feel about each of these activities? Are they different? The same? Should the reason for baring breasts have any effect on legislation of public nudity? What about genitals? Should they also be publicly showcased?
3 Are there any other diseases besides breast cancer around which social movements have emerged? If so, what are they? Is breast cancer different? How so? What do you make of the popularization of genetic explanations for breast cancer and the availability of BRCA1 testing used by Angelina Jolie?
4 In the United States, breast cancer—like other diseases—is stratified by race, with black women experiencing higher mortality rates. Discuss this in relation to your own geographic setting. Which women "get sicker and die quicker" from breast cancer? What are some explanations for this?
5 Discuss the reasons why environmental issues have been little researched with respect to breast cancer. What would be the consequences of discovering that breast cancer rates are causally related to toxins and pollution?

chapter 17

Genitals

Introduction

At first glance, the placement of this chapter in a section entitled "What Others See" might seem incongruous. Surely our genitals, those deeply private and sexualized body parts, are often kept hidden from view. Indeed, across cultures a variety of decency laws, religious teachings, and cultural norms instruct us regarding the social regulation of our genitals (and, for women, our breasts). We are taught from early ages to cover, conceal, wipe, groom, wash, and control our genitals.

In our own personal lives as mothers, we are daily on the "front lines" of transmitting at least some of these (U.S., middle-class, white) cultural norms to our children. However, at the same time, as feminists we are eager to provide our daughters with positive, progressive information that celebrates their sexual becoming and challenges dominant assumptions and practices. In our previous collaborations (e.g., Casper and Moore 2009), we have narrated how between us, we have given birth to and are mothering five daughters. As toddlers, tweens, and teens, our girls have examined, talked about, explored, celebrated, and narrated their vulvas, vaginas, and breasts along with their skin, hair, muscles, and other body parts.

Witnessing their varied journeys through socialization, including the internalization of the notion of "private parts" and "down there," has led to countless opportunities for us to interpret the interrelationship of female genital anatomy and misogyny in contemporary cultures, and in particular the United States where we live. In spite of our homosocial parenting situations (that is, we are women mothering girls), penises do not go unnoticed or invisible. Common curiosity, health class, popular media representations, and innuendo bring the penis into the individual imaginations of our daughters and conversations at our kitchen tables.

A sampling of the girls' questions over the years have included, for example: Why does it feel like milk when I touch my vagina? (2 year old) But why is my vulva private? (5 year old) What is a clitoris and do I have one? (6 year old) Do boys wish they had vaginas? (9 year old) How big do penises get? (10 year old) How does the penis get back to normal after an erection? (11 year old) What does sex feel like? (11 year old) Why don't we see as many penises as we see vaginas in the movies? (14 year old) And the ever-present question, when will I get my period?

And while we each have provided answers to our daughters that attempt to minimize potential body anxiety and shame, while also instilling social norms to keep them safe, these questions often lead to other questions and rarely are answered once and for all. Significantly, we have noticed that the older our daughters become, the less frequent, common, and plainly spoken these questions are, and instead childhood curiosity is supplanted by increasing layers of body humiliation, awkwardness, shame, and privacy.

This chapter offers a feminist sociological tour of human genitalia. We begin with an analysis of the "definition of human genitalia" and how, even in the defining of these body parts, we are engaging in the cultural work of creating normative social and anatomical categories. The ways in which we come to see male genitalia, in particular, are deeply related to prevailing ideas of phallocentrism and hegemonic masculinity that permeate the majority of cultures around the world.

Following this discussion, we turn to the variations of intersexuality and what has been made surgically, culturally, and politically of embodied differences. We take a closer look at male and female genital modifications, including circumcisions and cuttings. Then we offer a closer look at body modifications of the vulva—vajazzling, bikini waxes, and clitoral hood piercings. Also we highlight the foreskin restoration movement as a means of advocating for "intact" men and boys.

We conclude with a discussion of some of the ways these "private" yet quite public body parts—what performance artist Terri Kapsalis (1997) called "public privates"—are thought not to function as they should and the medical interventions suggested and used to correct perceived body "problems."

Locating human genitalia

As we have discussed throughout this book, *human anatomy* is the science of mapping the internal and external human body, most commonly through direct observation of cadavers. The topography of the body is then represented in books and other media through diagrams for training health care providers and the lay public about the form, function, and interconnection of body parts. Pelvic anatomy is where we find the genitals.

Genitals are the sex organs in the pelvic region of males and females and are typically classified into external genitalia, visible outside of the body, and internal genitalia, located inside the body.

In the case of female genitals, there is often a misuse of the word vagina to refer to the entire genital area. But the external female genitalia is called the vulva, which is made up of the labia (the folds of skin) and the clitoris, highly sensitive spongy tissue. The vagina is the internal genitalia and extends from the vaginal opening to the tip of the cervix, or the opening of the uterus. The uterus (sometimes called the womb), the ovaries, and fallopian tubes (also called oviducts) each participate in the female reproductive cycle of monthly egg release.

External male genitals are comprised of the penis and the scrotum, which contains the testes. The internal male genitals include the epididymis, a tube-like storage structure for sperm, the *vas deferens*, the sperm duct, and accessory glands including seminal vesicles, the prostate gland, and bulbourethral glands.

Phallocentrism

Anatomy is not a science devoid of bias. As Lisa Jean with sociologist Adele E. Clarke (1995) wrote in their extensive review of human genital anatomy: "In sum, we have examined an array of 20th century anatomy texts seeking both representativeness and range of variation. We found that by and large female sexual anatomy was given short shrift compared with that of the male both literally and figuratively." *Phallocentrism*, or the privileging of the male viewpoint or prerogative to support the domination of men over women, is prevalent in many cultures across historical time periods.

The phallus, a representation of the penis, is a model for both ancient and contemporary architectural structures. The Mayan phallic structures of Chichén Itzá in Mexico, stone phallic symbols recovered at Zimbabwean ruins, the Washington Monument in the District of Columbia, the Eiffel Tower in Paris, and most recently, the Swiss Re Building in London—all are phallic structures erected to celebrate masculine accomplishment and male power.

Preoccupation with the phallus is not only manifested in buildings. Linguist Deborah Cameron (1992) conducted a study of American college students to explore words and phrases that refer to the penis. She compiled a list of more than 140 terms, which she divided into categories, such as:

- titles of authority: "Kimosabe, his Excellency, your Majesty, the chief, the commissioner, the mayor, the judge" (370)

- personal authority: "scepter, rod of lordship, Excalibur, hammer of the gods" (370)
- tool: "screwdriver, drill, jackhammer, chisel, lawnmower, hedgetrimmer, and fuzzbuster" (371)
- weapon: "squirt gun, love pistol, passion rifle, lightsaber" (372).

Cameron concludes that the ways the penis is signified through language reinforces the underlying ideology that "the phallus must act, dominate, avenge itself on the female body. It is a symbol of authority to which we all must bow down. Its animal desires are uncontrollable; it has a life of its own" (1992: 373). Any reference to the vulnerable penis is notably absent. Yet philosopher Susan Bordo (1999: 64) has discussed personal and cultural anxiety surrounding penises:

> Most of our metaphors for the penis, as you will recall, actually turn it into some species of dildo: stiff torpedoes, wands, and rods that never get soft, always perform. These metaphors . . . may be a defense against fears of being too soft, physically and emotionally. But at the same time as these metaphors "defend" men as they joke with each other in bars—or more hatefully—act as a misogynist salve for past or imaginary humiliations, they also set men up for failure. For men don't really have torpedoes or rods or heroic avengers between their legs. They have penises. And penises, like the rest of the human body and unlike dildos, feel things.

Phallocentrism is related to another sociological concept: *hegemonic masculinity*. This term refers to the social, political, and economic domination of particular groups of men over others. Dominant men not only have social power; they also have cultural power. This combination is called hegemony.

Hegemonic men are economically successful, from racially and ethnically privileged groups, and visibly heterosexual; they are well educated or excel in their careers and work at the most prestigious and lucrative occupations. Some may be of poor, working class, or immigrant origins, but most have overcome their "humble" beginnings and have professional or managerial careers, pursue athletic goals, or enter politics. Their hegemonic status is produced and legitimated by a set of attributes highly valued, especially in the U.S. and Europe: whiteness, wealth, social position, and heterosexuality.

One of the paradoxical effects of patriarchal systems of domination is that even though men have dominance over women, men are not a monolithic group. In addition to being gendered, men are differently raced, differently abled, and differently classed. They can be gay, straight, bisexual, or asexual. In many cultures, these sexual and social differences are used to stratify groups of men into different categories of privilege and power.

Each man's ability to wield and benefit from patriarchal power is thus both different and often unequal. The ranking of culturally desirable male bodies mirrors their social standing.

Australian sociologist Raewyn Connell coined the term hegemonic masculinity as "the configuration of gender practice which embodies the currently accepted answer to the problem of the legitimation of patriarchy, which guarantees (or is taken to guarantee) the dominant position of men and subordination of women" (1995: 77). By describing the historical and social processes that put certain men at the top of the patriarchal hierarchy and make others complicit in supporting their position, Connell and other scholars in masculinity studies challenge the assumption that characteristics of masculinity arise from "natural" male bodies.

In contemporary Western society, the characteristics of hegemonic masculinity are embodied in an "ideal type" exhibiting the potential for physical power and violence, but acting with total rationality and control of emotion. It is an artificial image, based on cultural icons such as successful tycoons, movie actors, and sports figures, and does not reflect the reality of most men's actual behavior or circumstances. These icons, though, are powerful models, and they influence the behavior of men of many social classes and racial ethnic groups.

Intersex conditions

Intersexuality describes a variety of conditions in which an individual body's sexual, reproductive, and/or chromosomal anatomy does not correspond to medically and socially defined categories of male and female. For example, physiological/morphological sex (e.g., genitalia) may not "match" phenotypic sex. Rather than being perceived simply as a bodily difference, such mismatches—until very recently—have been deemed "abnormal" deviations and treated accordingly.

The term "hermaphrodite" was used for centuries to describe intersexuality, although it is not scientifically accurate and also has been heavily stigmatized. Derived from the Greek Hermaphroditus, nymph-son of Hermes and Aphrodite, the term implies the presence of both "male" and "female" in the same body. However, most intersex conditions are more complicated than this simple binary. Historically, "hermaphrodites" were fodder for legend and also were displayed in circuses as "freaks of nature" (Thomson 1996).

The intersex rights movement, most visibly represented by the now-defunct Intersex Society of North America founded by Cheryl Chase, helped to change social perceptions of intersexed individuals as well as medical practices (Karkazis 2008). The American Academy of Pediatrics had long considered the birth of an intersex child to be a "social emergency," and recommended surgical "treatment" of newborns to bring genitalia in line with other markers such as appearance and chromosomal makeup (Preves 2003). This position dovetailed with parents' anxieties and fears about raising a child born "different."

Intersex rights advocates argued that such interventions violated the rights of children, who are unable to give informed consent to surgery, and also directly contradicted the Hippocratic Oath to "first do no harm." ISNA, for example, advanced the notion that intersexuality is a problem of stigma, not gender, and that professional mental health and public education are the solution, not surgical intervention that cannot be reversed and may pose harmful consequences, such as problems with sexual function (Creighton and Minto 2001).

And though intersex rights advocates were often perceived as positing a "third gender" as a solution, ISNA was clear that children should be assigned a gender of "boy" or "girl" based on the best evidence, but without surgically removing body parts (except in life-threatening situations).

ISNA closed its doors in 2008, just after the Accord Alliance was born. Simultaneously, the language of intersex was replaced by some groups (including ISNA) with *disorders of sexual development* (DSD). Two handbooks were published to guide clinical practice and parental decision-making: Clinical *Guidelines for the Management of Disorders of Sex Development in Childhood* and a *Handbook for Parents*. While some advocates saw this as an advance in efforts to reform interventionist medical practice, others saw it as a capitulation to medicalized definitions of intersex (Feder and Karkazis 2008).

Although definitions of intersex remain contested, several developments in human rights law and policy have shifted the conversation about intersex to the rights of the child. In 1999, the Constitutional Court of Colombia recognized people with intersex conditions as a minority group entitled to protection against discrimination, and found that surgical interventions for intersex constitute a violation of autonomy and bodily integrity.

In 2005, San Francisco declared intersex a human rights issue, finding that it is "unethical" to surgically alter a child's genitals and that "normalizing" interventions are a violation of bodily integrity. This marked the first time that a government entity in the United States, in this case a municipality, deemed intersex a matter of human rights.

In New Zealand, the Human Rights Commission convened a series of roundtables to explore the human rights issues of intersexed individuals. In 2012, Europe's first intersex human rights organization, OIIEurope, was founded as an "umbrella" organization, joining Organization Intersex International (OII) World and numerous other national/regional NGOs including OII United Kingdom and OII USA.

Female genital cutting

Female genital cutting (FGC), sometimes called *female genital mutilation* (FGM), is the practice of intentionally altering female genital organs for nonmedical reasons and is usually performed on girls between the ages of infancy and 15. The most typical types of cutting include clitoridectomy, the partial or complete removal of the clitoris; excision, removal of the clitoris and labia minora and sometimes the labia majora; and infibulation, the narrowing of the vaginal opening.

FGC is practiced primarily in central Africa, southern Sahara, and regions of the Middle East; it is most prevalent in Egypt and Ethiopia. The World Health Organization estimates that, worldwide, between 100–140 million girls have been genitally altered and describes the practice as a violation of the human rights of girls and women. Cutting practices can lead to infections and medical complications, sexual dysfunction, adverse obstetrical outcomes (including fistula), and death.

Debates about FGC/FGM illustrate tensions between *ethnocentrism*, or using one's own culture to judge others, and *cultural relativism*, or trying to appreciate other groups' ways of life in context, without judging them as superior or inferior to one's own culture. In the case of FGC, there is a history of women and men from the Global North decrying FGC as "barbaric" and cruel without understanding the social, cultural and religious conventions of some of these practices. Critics of FGC typically prefer the more normative/moralistic language of mutilation rather than the more generic term cutting.

FGC is reportedly practiced in some families as a way to demonstrate love and care for their daughters, as a way of making girls marriage-ready, and not branding them with the stigma of being uncut. Furthermore, critics often do not examine similar types of sexist or discriminatory practices within their own cultures. For example, some feminist critics of FGC in the United States and Europe infrequently address practices of cosmetic genital surgery, such as labioplasty or vaginal tightening, prevalent in their own countries.

On the other hand, there is a profound danger in excusing FGC as something "other" cultures do without critically examining these practices as deeply lodged within systems of sex discrimination, male privilege, and heterosexism. Adopting the veil of "traditional practices" to avoid criticizing cutting fails to allow for critical examination of FGC practices, and in particular, of the obvious gendered dynamics.

Women in countries where FGC is prevalent continue to speak out against such practices in demonstrations, such as the one pictured below. Here we see local women (not activists from the West) appealing to religious leaders in their own villages and working painstakingly through

FIGURE 17.1 Senegal genital cutting
© Christian Aslund
Representatives from 210 villages meeting up for demonstration against child and enforced marriage and female genital mutilation.

social customs to suggest more woman-friendly alternatives to the practices of FGC.

The feminist politics of genital cutting are extraordinarily complex. Virginia Braun, a New Zealand psychologist, suggests that we must consider if *any* level of FGC is acceptable under any circumstances. She sees FGC as existing along a continuum from practices in African and Middle Eastern countries to those commonly found in Global North countries (2012).

For example, female genital cosmetic surgery, including hymenoplasty, labiaplasty, "G-spot" amplification, and vaginal "rejuvenation," alters vulval morphology but is considered elective surgery and not medically indicated. These interventions are seen as enhancements within carefully proscribed cultural paradigms of beauty and desirability. The same, however, could be said for FGC in more "traditional" cultures.

A group of researchers have studied the marketing of these elective procedures through online advertisements and found that both the quality and quantity of clinical information on sites is poor and that there is a distortion of information about these procedures (Liao, Taghinejadi, and Creighton 2012). The American Congress of Obstetricians and Gynecologists advises against cosmetic vaginal procedures since they are not deemed safe or effective.

It must be noted that neither FGC nor "elective" enhancement surgeries are done for life-saving or medically indicated reasons.

A CLOSER LOOK

Brazilian waxing, vajazzling, and female genital piercing

Similar to other practices of transforming body parts to be congruent with social norms of appearance (hairstyles, makeup application, hair dyes), female genitals are often modified in line with community norms. Procedures, much less invasive than FGC, include other methods that some women and girls use to enhance the appearance and appeal of their vulvas and labia.

A Brazilian wax is the removal of pubic hair from the pelvic area, often from the belly to the buttocks using hot wax and strips of gauze to pull the hair from the follicle. These waxes generally run between US$60–75 per visit and are typically undertaken every 3–6 weeks to maintain a "clean" result. The final product is a completely hairless vulva evocative of a young girl's prepubescent body.

Vajazzling, a term coined around 2010, is the practice of applying temporary glitter or jewels with adhesive to a woman's pubic area for aesthetic purposes. Women can either purchase a do-it-yourself (DIY) crystal bikini kit or visit a salon and have the jewels applied by an aesthetician (between US$50–$200 depending on the salon).

Female genital piercing, most commonly clitoral hood and labial piercings, involves the insertion of jewelry—usually stainless steel studs, barbells, or hoops—into the flesh of the female genital area. Piercing is done both for enhanced sexual sensation and visual appeal. Generally, each piercing runs between US$60–150 and requires some aftercare to reduce risk of infection. Penises may also be pierced. Evocative of our discussion earlier about nicknames for the phallus, one type of piercing is called a Prince Albert.

The "grooming" of female (and male) external genitalia requires considerable financial investment and might be considered a distinctive kind of "presentation of self" (Goffman 1959). While genital "enhancement" is most typically seen as conforming to cultural ideals of beauty (including femininity and masculinity), in some contexts—such as very conservative communities and families or highly religious contexts—a practice such as genital piercing might be viewed as rebellious.

Male circumcision

It is worth considering that the vulnerability and anxiety some men feel about their penises may be related to their earliest childhood experiences, even experiences in infancy. Before a baby's tiny body has been gendered by the bombardment of unattainable media images of the "perfect" male form, many male babies are implicated in ongoing debates about a different kind of genital cutting: male circumcision.

Male circumcision, or removal of the foreskin of the penis, is most commonly a sign of ethnic or religious identity; it is practiced in Muslim and Jewish religions as largely obligatory for male children. According to the World Health Organization, globally approximately one in three males is circumcised. However, in some societies, circumcision of newborns in hospitals for *nonreligious* reasons became routine in the 20th century. Owing to changes in beliefs systems, trends in biomedical research, and social movements raising awareness about circumcision, male circumcision rates have varied widely over the past 50 years (Carpenter 2010).

The most recent data reveal that the overall circumcision rate in the United States for newborn males prior to release from the hospital was 56 percent in 2005 (Circumcision Reference Library 2009). However, a representative nationwide survey of adult males in the U.S. found an overall rate of 79 percent, with rates varying by race and ethnicity: 88 percent in nonHispanic whites, 73 percent in blacks, 42 percent in Mexican-Americans, and 50 percent in "others" (Xu and others 2007). Non-ritual circumcision is uncommon in Asia, South and Central America, and most of Europe—with a global rate of approximately 34 percent.

A review by the American Academy of Pediatrics of benefits of circumcision found that the rate of urinary tract infections in the first year of life was seven to 14 of 1,000 uncircumcised male infants and one to two of 1,000 circumcised male infants; however, the absolute risk of an uncircumcised male infant developing a urinary tract infection in the first year of life is less than one percent (AAP 1999). Neonatal circumcision offers some protection from cancer of the penis in later life; here, the overall risk is even lower—nine to ten cases a year for every one million men.

In contrast, the risk of the circumcision itself (accidental injury, hemorrhaging, and infection) was one out of 476 in one study (Christakis and others 2000). In 1999, the AAP concluded that the incidence of urinary tract infection and penile cancer in the United States is too low to warrant routine circumcision of male infants and now recommends that parents be informed of the potential benefits *and* risks of circumcision and that it is a strictly elective procedure.

In 2012, the AAP updated its position, suggesting that benefits outweigh risks but not enough "to recommend universal circumcision." The update was met with considerable controversy by critics who contend that male circumcision, like FGC, is a human rights violation.

A different picture of the benefits of circumcision presents in sub-Saharan Africa, where circumcision has been shown to be highly effective in protecting men from the risk of HIV infection through heterosexual transmission (Lemle 2005; O'Farrell and Egger 2000; Weiss et al. 2000). There may be several reasons for this protection. First, the foreskin is vulnerable to lesions, tears, and ulceration, which increase the likelihood of transmission of HIV and other sexually transmitted diseases. Second, the foreskin contains a high density of cells that may be the primary target for HIV infection (Szabo and Short 2000).

This positive effect of circumcision has been touted as a rationale for public health officials in the United States and Europe to consider promoting routine circumcision for all baby boys. Proposed policy shifts have led to controversy. In Europe, for example, decreasing circumcision rates are seen not in terms of public health, but as a reflection of increasing secularization. In the United States, "members of Intact America, a group that opposes newborn circumcision, have rented mobile billboards that will drive around Atlanta carrying their message that 'circumcising babies doesn't prevent H.I.V.'" (Rabin 2009).

An especially volatile issue in the circumcision debate is whether removal of the foreskin diminishes sexual sensation. In the 19th century, circumcision (and clitorectomy in women) was recommended as a deterrent to masturbation, which led activists to argue that removal of the foreskin must deaden sexual feeling (Zoske 1998). However, data from 1410 U.S. men aged 18 to 59 who participated in the 1992 National Health and Social Life Survey found that self-identified circumcised men had only a slightly lower risk of experiencing sexual dysfunction, such as inability to have orgasms, especially when older (Laumann et al. 1997). The survey also found that circumcised men engaged in more varied sexual practices and masturbated more, but the variation across ethnic groups suggested that differences were due to social factors, which correlate with likelihood of being circumcised.

It is unlikely there will ever be fully objective data on differences in sexual pleasure given and received by men with and without foreskins, since these evaluations are so subjective and research on sexuality so contested.

Given widespread cultural preoccupation with the penis, the ritual practice of circumcision in male newborns or at puberty seems puzzling. However, for ritual circumcision, it is precisely the preciousness of the penis that underlies the rite. Male circumcision ceremonies indicate a father's loyalty to his lineage elders—"visible public evidence that the head of a family unit of their lineage is willing to trust others with his and his family's most valuable political asset, his son's penis" (Erickson-Paige and Paige 1981: 147).

As recently as 2012, controversy continues to swirl around male circumcision globally, revealing tensions between religion and secularization, human rights and traditional practices, cultural values and laws, and history and the present. Several countries—including Germany, Austria, Norway, and Denmark—saw attempts to outlaw circumcision.

While the arguments often focus on questions of bodily integrity, anti-circumcision advocacy is congruent with anti-Jewish and anti-Muslim sentiment, leading some critics to suggest that the circumcision debate may sometimes be about race, ethnicity, and religion. Those who oppose circumcision point out that infants cannot consent to the procedure (as with intersex "treatment"), and perhaps the solution is to "let boys decide for themselves" when they are old enough to understand the risks and benefits of the procedure.

A CLOSER LOOK

The foreskin restoration movement

The practice of foreskin restoration involves using weights, elastics, or cones to cover the glans penis with a double sheath of retractable tissue. In general these nonsurgical options, which rely on tension through pulling and stretching, are far more common than the less successful surgical procedure (Griffiths, Bigelow, and Loewen 2010).

Men self-report wanting to restore their foreskin for appearance reasons, for greater sexual stimulation, and/or for relief of discomfort due to oversensitivity of the penis. In addition, some men report feeling low levels of body satisfaction about their circumcisions.

United States-based groups such as National Organization of Restoring Men, Foregen, and the National Organization of Circumcision Information Resource Centers advocate for genital integrity for men through foreskin restoration and regeneration research for men who are already circumcised. They work to raise awareness about circumcisions for parents considering the procedure.

What's wrong with women's sexuality?

Female sexual dysfunction is a term used to describe recurrent or persistent problems with sexual response or sexual desire in women. Obviously, the term raises the political, ethical question: for whom is women's sexual response perceived to be a problem? Sociologist and bioethicist Jennifer

Fishman (2004), among other critics, has suggested that female sexual dysfunction was manufactured by the pharmaceutical industry intent on seeking new markets for drugs.

Since overwhelmingly in heterosexual encounters, women are the "receptive" partners who facilitate male orgasm, their disinterest in sex might be a problem for their "insertive" partners. These "problems" include concerns with low sex drive, vaginal lubrication, painful sexual intercourse, and/or being nonorgasmic. Because many women self-report experiencing all these symptoms at some point in their sexual lives, it is difficult to know what, precisely, is "dysfunctional" about them. For example, in 1999, the *Journal of the American Medical Association* claimed that perhaps 43 percent of American women suffered from FDS.

Like Fishman, journalist Ray Moynihan (2003) has suggested that researchers directly linked to pharmaceutical companies created the term "female sexual dysfunction" based on limited research as a means of constructing new markets for diagnosis and treatment of emerging "disorders." Related terms include, for example, "female sexual arousal disorder" and "hypoactive sexual desire disorder." This process is known as *disease mongering*, or the creation of pathological descriptions of normal bodily functioning for pharmaceutical profit (Payer 1992), which is related to *biomedicalization* (Clarke et al. 2009).

Understanding one's own sexual response is often riddled with shame and taboos; and in societies that oppress women, it is deeply challenging to develop the self-knowledge for sexual exploration. Because there is scant research on women's sexual function and because phallocentrism is so prevalent in much human sexuality research, it is difficult to establish a baseline of what would be considered "healthy" female sexuality.

It is common for people to ask, "What is wrong with me?" when it comes to experiencing and expressing a range of sexual desires. As psychologist Leonore Tiefer explains, individuals might desire female sexual dysfunction to be seen as a medical problem: "The public finds medicalization attractive because the notion of simple but scientific solutions fits in with a general cultural overinvestment in biological explanations and interventions, and promises to bypass sexual embarrassment, ignorance, and anxiety" (2006: 439).

What's wrong with men's sexuality?

Sociologically, we are curious as to how men have become so concerned with presenting a certain image of their penises. We suggest that a focus on sexual performance as a proof of masculinity is something that is learned by boys and reinforced in men in Western cultures through multiple sources. From pornography to sexology research to locker room humor, erotic images and human sexual responses are defined as almost exclusively phallocentric. The sexual act is understood to be penetration (vaginal, oral, or anal) and the goal is male orgasm and ejaculation. Anything less is tantamount to an embarrassing failure; repeated experiences end up with a diagnosis of "erectile dysfunction"—impotence (Tiefer 1994).

Recent innovations in pharmaceuticals, as well as a long history of penis-improvement gadgets, also construct male sexuality in this very limited way. Simply checking your email spam folder demonstrates how we are bombarded with constant efforts to market and sell penile enhancement.

Erectile dysfunction (ED) is defined as the inability to achieve or maintain an erection for sexual performance. Urologists predict that due in part to growing life expectancy, the prevalence of ED will increase substantially through 2025; these urologists suggest that rising rates of ED will place high demand on underfunded public health systems for treatment (Aytac, McKinlay, and Krane 1999).

In situations of limited health care funding, how do governments prioritize the treatment of diseases and disorders? Can only men able to purchase ED pharmaceuticals "get hard"? How can nations justify providing ED drugs when they do not make basic contraceptives for women accessible and affordable? How do markets and resources for ED drugs compare to those for female sexual dysfunction? Who gets to benefit from new pharmaceuticals?

Men's vulnerability about their sexual functioning or "prowess" goes a long way toward explaining the enormous popularity of Viagra and similar medications, and their use by men whose only medical indication is fear of sexual failure (Loe 2004). In order to create less shame about threats to masculinity due to impotence, drug companies select dominant (usually white) males to be spokesmen for their products, for example, former U.S. Senator Bob Dole and former Chicago Bears football coach Mike Ditka. Despite their soft-focus imagery, advertisements for Viagra, Cialis, Levitra, and similar pharmaceuticals are not selling the sensuality and emotions of sexual relationships, but a "magic bullet" method to achieve otherwise unattainable standards of the perfectly functioning penis and the always successful sexual performance.

Conclusion

This chapter has examined varied and contested social and cultural meanings and practices surrounding human genitalia. While often considered "private," penises and vulvas are actually quite public and visible body parts in which we invest cultural meanings about sex, gender, and power.

Our genitals are examined, displayed, sliced, probed, decorated, erased, and otherwise manipulated, by others and ourselves, in the service of broader social, cultural, and political aims. From self-enhancements such as piercing to the potentially violent surgical interventions of intersex "treatment," FGC, and male circumcision, genitals are very much a part of social life.

The ultimate irony is that these deeply private and often unseen body parts have a starring role in everything from family life and intimate relations to medical practice and human rights.

See also

Fluids; Reproduction; Hair; Breasts

Key terms

Human anatomy, phallocentrism, hegemonic masculinity, male circumcision, foreskin restoration, female genital cutting, female genital mutilation, cultural relativism, ethnocentrism, intersexuality, disorders of sex development, vajazzling, female genital piercing, impotence, vaginal dryness, disease mongering

Discussion questions

1 Considering that human genitals are considered private parts, how do we come to know what they are? Who are some of the people who taught you about your genitals? What are some of the things you remember "learning"? Where are the most common places we see human genitals? Who is responsible for narrating what these body parts do? Discuss some common misconceptions or anxieties about human genitals in your culture.

2 What are some of the most prevalent examples of phallocentrism in everyday life? Collect and bring in a few examples of images that represent phallocentrism. What do you think these images convey? For example, locate an image of an advertisement for ED treatments. What are the implications of the advertisement?

3 Considering your own culture and society, why do you think non-normative genitalia, such as those in children with intersex conditions, are so threatening? How do gender politics and sexual mores

influence what we think about genitals? What are some alternatives to surgical intervention to addressing intersex conditions?

4 Disease mongering is the process of creating and/or exaggerating physical or psychological differences to produce a medical problem, particularly when engineered by someone who will benefit from the treatment of the new "disease." Why are genitals, as body parts, vulnerable to disease mongering? Discuss who benefits when genitals become "sick."

5 Discuss the types of genital surgeries, modifications, and adjustments individuals make to bring their genitals into a more culturally desirable aesthetic. What does this range of procedures say about our relationships to our genitals? What do the procedures say about perceived gendered roles and gendered performances?

References

Ackery, Alun, Charles Tator, and Andrei Krassioukov. 2004. "A Global Perspective on Spinal Cord Injury Epidemiology." *Journal of Neurotrauma* 21(10): 1355–70.

Ahmed, S. and S. A. Holtz. 2007. "Social and Economic Consequences of Obstetric Fistula: Life Changed Forever?" *International Journal of Gynecology and Obstetrics* 99: S10–S15.

Ahmed, Saifuddin, Qingfeng Li, Li Liu, and Amy O. Tsui. 2012. "Maternal Deaths Averted by Contraceptive Use: An Analysis of 172 Countries." *The Lancet* 380(9837) 111–25.

Aidoo, Michael et al. 2002. "Preventive Effects of the Sickle Cell Gene Against Malaria Morbidity and Mortality." *The Lancet* 359(9314): 1311–12.

Akashi, Yoshihiro J., Holger M. Nef, Helga Möllman, and Takashi Ueyama. 2010. "Stress Cardiomyopathy." *Annual Review of Medicine* 61: 271–86.

Albertini, Richard, Michael Bird, Nancy Doerrer, Larry Needham, Steven Robison, Linda Sheldon, and Harold Zenick. 2006. "The Use of Biomonitoring Data in Exposure and Human Health Risk Assessments." *Environmental Health Perspectives* 114(11): 1755–62.

Alexander, Michelle. 2012. *The New Jim Crow: Mass Incarceration in the Age of Colorblindness*. New York: The New Press.

Alio, Amina P., Philip N. Nana, and Hamisu M. Salihu. 2009. "Spousal Violence and Potentially Preventable Single and Recurrent Spontaneous Fetal Loss in an African Setting: Cross-Sectional Study." *The Lancet* 373(9660): 318–24.

Allen, Monica J., Michelle E. Powers, K. Scott Gronowski, and Ann M. Gronowski. 2010. "Human Tissue Ownership and Use in Research: What Laboratorians and Researchers Should Know." *Clinical Chemistry* 56(11): 1675–82.

Almeling, Rene. 2011. *Sex Cells: The Medical Market for Eggs and Sperm*. Berkeley: University of California Press.

Althuis, Michelle D. et al. 2005. "Global Trends in Breast Cancer Incidence and Mortality 1973–1997." *International Journal of Epidemiology* 34(2): 405–12.

Amato, Gennaro, et al. 2013. "Climate Change, Air Pollution and Extreme Weather Events Leading to Increasing Prevalence of Allergic Respiratory Diseases." *Multidisciplinary Respiratory Medicine*. 8(1): 12.

American Diabetes Association. 2013. "Standards of Medical Care in Diabetes 2013." *Diabetes Care* 36 Suppl 1: S11–S66.

Andersen, Peter. 2007. *Nonverbal Communication: Forms and Functions*. 2nd edition. Long Grove, IL: Waveland Press.

Anderson, Benjamin O. and R. Jakesz. 2008. "Breast Cancer Issues in Developing Countries: An Overview of the Breast Health Global Initiative." *World Journal of Surgery* 32(12): 2578–85.

Anderson, Benjamin O. et al. 2011. "Optimisation of Breast Cancer Management in Low-Resource and Middle-Resource Countries: Executive Summary of the Breast Health Global Initiative Consensus, 2010." *The Lancet Oncology* 12(4): 387–98.

Anderson, Warwick. 2003. *The Cultivation of Whiteness: Science, Health, and Racial Destiny in Australia*. New York: Basic Books.

Andreff, Wladimir. 2006. "Sports in Developing Countries." In Andreff, Wladimir and S. Szymanski (eds.). *Handbook on the Economics of Sport*. Cheltenham: Edward Elgar, 308–15.

Angier, Natalie. 1999. *Woman: An Intimate Geography*. New York: Anchor Books.

Annas, George J. and Michael A. Grodin, eds. 1995. *The Nazi Doctors and the Nuremberg Code: Human Rights in Human Experimentation*. New York: Oxford University Press.

Appadurai, Arjun. 1996. *Modernity At Large: Cultural Dimensions of Globalization*. Minneapolis: University of Minnesota Press.

Aronson, Josh [director]. 2000. *Sound and Fury*. Produced by Roger Weisberg. Distributed by Aronson Film Associates, Inc. and Public Policy Productions, Inc., in association with Thirteen/WNET New York and Channel 4 (UK).

Ashikari, Mikiko. 2003. "Urban Middle-Class Japanese Women and Their White Faces: Gender, Ideology and Representation." *Ethos* 31(1): 3–37.

Atkin, Karl and Waqar I. U. Ahmad. 1998. "Genetic Screening and Haemoglobinopathies." *Social Science and Medicine* 46(3): 445–58.

Atwood, Kimball C. 2004. "Naturopathy, Pseudoscience, and Medicine: Myths and Fallacies vs Truth." *Medscape* 6(1): e53.

Aytac, I. A., J. B. McKinlay, and R. J. Krane. 1999. "The Likely Worldwide Increase in Erectile Dysfunction between 1995 and 2025 and Some Possible Policy Consequences." *British Journal of Urology International* 84: 50–56.

Baddour, J. A., K. Sousounis, and P. A. Tsonis. 2012. "Organ Repair and Regeneration: An Overview." *Birth Defects Research, Part C: Embryo Today* 96(1): 1–29.

Bainbridge, William Sims. 2003. "Religious Opposition to Cloning." *Journal of Evolution and Technology* 13. http://www.jetpress.org/volume13/bainbridge.html.

Baker, Monya. 2012. "Biorepositories: Building Better Biobanks." *Nature* 486: 141–46.

Banks, Marcus and Jay Ruby, eds. 2011. *Made To Be Seen: Perspectives on the History of Visual Anthropology*. Chicago: University of Chicago Press.

Barad, Karen. 2007. *Meeting the Universe Halfway: Quantum Physics and the Entanglement of Matter*. Durham, NC: Duke University Press.

Barndt, Deborah. 2008. *Tangled Routes: Women, Work, and Globalization on the Tomato Trail*. 2nd edition. Lanham, MD: Rowman & Littlefield.

Baudrillard, Jean. 1994. *Simulacra and Simulation*. Ann Arbor: University of Michigan Press.

Bauman, H.-Dirksen L. and Joseph J. Murray. "Deaf Studies in the 21st Century: 'Deaf-Gain' and the Future of Human Diversity." pp. 246–62 in *The Disability Studies Reader*, edited by Lennard J. Davis. 4th edition. New York: Routledge.

Bauman, Zygmunt. 2005. *Liquid Life*. Cambridge: Polity Press.

Bell, Susan E. 2009. *DES Daughters, Embodied Knowledge, and the Transformation of Women's Health Politics in the Late Twentieth Century*. Philadelphia: Temple University Press.

Benn, Peter and Audrey Chapman. 2010. "Ethical Challenges in Providing Noninvasive Prenatal Diagnosis." *Current Opinions in Obstetrics and Gynecology*. 22: 128–34.

Bennett, I. C., M. Gattas, and B. T. Teh. 1999. "The Genetic Basis of Breast Cancer and Its Clinical Implications." *ANZ Journal of Surgery* 69(2): 95–105.

Bennett, Jane. 2010. *Vibrant Matter: A Political Ecology of Things*. Durham, NC: Duke University Press.

Berger, John. 1972. *Ways of Seeing*. London: British Broadcasting Corporation.

Berger, Peter L. and Thomas Luckmann. 1967. *The Social Construction of Reality: A Treatise in the Sociology of Knowledge*. New York: Anchor Books.

Bergman, Ake, Jerrold J. Heindel, Susan Jobling, Karen A. Kidd and R. Thomas Zoeller, (eds). 2012. *State of the Science of Endocrine Disrupting Chemicals*. Geneva: World Health Organization.

Bering, Jesse. 2009. "Would Your Clone Have Its Own Soul, Or Be a Soulless Version of You?" Scientific American, February 13. http://www.scientificamerican.com/article.cfm?id=would-your-clone-have-a-soul.

Bermúdez-Rattoni, Federico, ed. 2007. *Neural Plasticity and Memory: From Genes to Brain Imaging*. Boca Raton, FL: CRC Press.

Bernheimer, H., W. Birkmayer, O. Hornykiewicz, K. Jellinger, and F. Seitelberger. 1973. "Brain Dopamine and the Syndromes of Parkinson and Huntington: Clinical, Morphological, and Neurochemical Correlations." *Journal of the Neurological Sciences* 20(4): 415–55.

Berry, Esther. 2008. "The Zombie Commodity: Hair and the Politics of Globalization." *Postcolonial Studies* 11(1): 63–84.

Betrán, A. P., M. Merialdi, J. A. Lauer, W. Bing-Shun, J. Thomas, P. Van Look, and M. Wagner. 2007. "Rates of Cesarean Section: Analysis of Global, Regional, and National Estimates." *Pediatric and Perinatal Epidemiology* 21(2): 98–113.

Blank, Hanna. 2012. *Straight: The Surprisingly Short History of Heterosexuality*. Boston, MA: Beacon Press.

Bliss, Catherine. 2012. *Race Decoded: The Genomic Fight for Social Justice*. Palo Alto, CA: Stanford University Press.

Blum, Linda M. 2000. *At the Breast: Ideologies of Breastfeeding and Motherhood in the Contemporary United States*. Boston, MA: Beacon Press.

Bock, B. C., J. L. Fava, R. Gaskins, K. M. Morrow, D. M. Williams, E. Jennings, and B. H. Marcus, 2012. "Yoga as a Complementary Treatment for Smoking Cessation in Women." *Journal of Women's Health* 21(2): 240–48.

Boehmer, Ulrike. 2002. "Twenty Years of Public Health Research: Inclusion of Lesbian, Gay, Bisexual, and Transgender Populations." *American Journal of Public Health* 92(7): 1125–30.

Bolfert, Kristin, "Antoni van Leeuwenhoek". Embryo Project Encyclopedia (2007-11-01). ISSN: 1940-5030. http://embryo.asu.edu/handle/10776/1704.

Bonham, Vence L., Esther Warshauer-Baker, and Francis S. Collins. 2005. "Race and Ethnicity in the Genome Era: The Complexity of the Constructs." *American Psychologist* 60(1): 9–15.

Bookman, Milica Z. and Karla R. Bookman. 2007. *Medical Tourism in Developing Countries*. New York: Palgrave Macmillan.

Bordo, Susan R. 2000. *The Male Body: A New Look at Men in Public and Private*. New York: Farrar Straus Giroux.

Borg, Johan, Anna Lindstrom, and Stig Larsson. 2009. "Assistive Technology in Developing Countries: National and International Responsibilities to Implement the Convention on the Rights of Persons with Disabilities." *The Lancet* 374: 1863–65.

Boseley, Sarah. 2012. "Stem Cell Scientists Urge MEPs Not to Cut Funding for Human Embryo Research." *The Guardian*, June 15.

Bourdieu, Pierre. 1984. *Distinction: A Social Critique of the Judgement of Taste*. Cambridge, MA: Harvard University Press.

Bowker, Geoffrey C. and Susan Leigh Star. 2000. *Sorting Things Out: Classification and Its Consequences*. Cambridge, MA: MIT Press.

Braithwaite, Jeanine and Daniel Mont. 2009. "Disability and Poverty: A Survey of World Bank Poverty Assessments and Implications." *ALTER/European Journal of Disability Research* 3(3): 219–32.

Braman, Sidney. 2006. "The Global Burden of Asthma." *Chest* 130(1): 4S.

Braun, Virginia. 2010. "Female Genital Cosmetic Surgery: A Critical Review of Current Knowledge and Contemporary Debates." *Journal of Women's Health* 19(7): 1393–407.

Braun, Virginia. 2012. "Female Genital Cutting around the Globe: A Matter of Reproductive Justice." In Joan Chrisler eds, *Reproductive Justice*. Santa Barbara, CA: ABC Clio Press, 29–45.

Brebner, Joanne L. et al. 2009. "Dude Looks Like a Lady: Exploring the Malleability of Person Categorization." *European Journal of Social Psychology* 39: 109–19.

Briggs, Laura. 2002. *Reproducing Empire: Race, Sex, Science, and U.S. Imperialism in Puerto Rico*. Berkeley: University of California Press.

Briggs, Laura. 2012. *Somebody's Children: The Politics of Transracial and Transnational Adoption*. Durham, NC: Duke University Press.

Brodal, Per. 2010. *The Central Nervous System*. 4th edition. New York: Oxford University Press.

Brown, Phil. 1992. "Popular Epidemiology and Toxic Waste Contamination: Lay and Professional Ways of Knowing." *Journal of Health and Social Behavior* 33(September): 267–81.

Brown, Stephen. 2008. "Miscarriage and Its Associations." *Seminars in Reproductive Medicine* 26(5): 391–400.

Browne, Kath. 2004. "Genderism and the Bathroom Problem: (Re)materialising Sexed Sites (Re)creating Sexed Bodies." *Gender, Place and Culture* 11(3): 331–46.

Browner, Carole H. and Carolyn F. Sargent, eds. 2011. *Reproduction, Globalization, and the State: New Theoretical and Ethnographic Perspectives*. Durham, NC: Duke University Press.

Brumberg, Joan Jacobs. 1998. *The Body Project: An Intimate History of American Girls*. New York: Vintage.

Bruner, Ed. 1996. "Tourism in the Balinese Borderzone." In *Displacement, Diaspora, and Geographies of Identity*, edited by S. Lavie and T. Swedenberg. Durham & London: Duke University Press.

Budiani-Saberi, D. A. and F. L. Delmonico. 2008. "Organ Trafficking and Transplant Tourism: A Commentary on the Global Realities." *American Journal of Transplantation* 8(5): 925–29.

Bujalkova, M. 2007. "Birth Control in Antiquity." *Bratislava Medical Journal* 108(3): 163–66.

Bullard, Robert D. 1990. *Dumping in Dixie: Race, Class, and Environmental Quality*. Boulder, CO: Westview.

Bunn, H. Franklin. 1997. "Pathogenesis and Treatment of Sickle Cell Disease." *New England Journal of Medicine* 337: 762–69.

Burgard, Deb. (2009). "What Is Health at Every Size?" In Rothbum, Esther and Solovay, Sondra (eds.) *The Fat Studies Reader* (pp. 42–53). New York, NY: New York University Press.

Burke, Willie, Beth Tarini, Nancy Press, and James Evans. 2011. "Genetic Screenings." *Epidemiologic Reviews* 33(1): 148–64.

Butler, Judith. 1990. *Gender Trouble: Feminism and the Subversion of Identity*. New York: Routledge.

Butler, Judith. 1993. *Bodies that Matter: On the Discursive Limits of "Sex."* New York: Routledge.

Cameron, Deborah. 1992. "Naming of Parts: Gender, Culture and Terms for the Penis among American College Students." *American Speech* 67: 367–82.

Caplan, Jane and John Torpey. 2001. *Documenting Individual Identity: The Development of State Practices in the Modern World*. Princeton: Princeton University Press.

Carney, Scott M. 2011. *The Red Market: On the Trail of the World's Organ Brokers, Bone Thieves, Blood Farmers, and Child Traffickers*. New York: William Morrow.
Carney, Scott. 2010. "The Temple of Do." *Mother Jones* March/April.
Carolan, Michael. 2010. *The Sociology of Food and Agriculture*. New York: Routledge.
Carpenter, Laura M. 2010. "On Remedicalization: Male Circumcision in the United States and Great Britain." *Sociology of Health and Illness* 32(4): 613–30.
Carroll, Katherine and K. Herrmann. 2012. "Introducing Donor Human Milk to the NICU: Lessons for Australia." *Breastfeeding Review* 20(3): 19–26.
Carter, Rita. 2009. *The Human Brain Book*. London: DK Adult.
Casals-Casas, C. and Desvergne, B. (2011). "Endocrine Disruptors: From Endocrine to Metabolic Disruption." *Annual review of physiology* 73: 135–62.
Casper, Monica J. 1994. "At the Margins of Humanity: Fetal Positions in Science and Medicine." *Science, Technology and Human Values* 19(3): 307–23.
Casper, Monica J. 1998. *The Making of the Unborn Patient: A Social Anatomy of Fetal Surgery*. New Brunswick, NJ: Rutgers University Press.
Casper, Monica J. (ed.). 2003. *Synthetic Planet: Chemical Politics and the Hazards of Modern Life*. New York: Routledge.
Casper, Monica J and Lisa Jean Moore. 2009. *Missing Bodies: The Politics of Visibility*. New York: New York University Press.
Cavanagh, Sheila. 2010. *Queering Bathrooms: Gender, Sexuality, and the Hygienic Imagination*. Toronto, ON: University of Toronto Press.
Centers for Disease Control. 2008. Newborn Screening Laboratory Bulletin. Accessed on May 23, 2013. http://www.cdc.gov/nbslabbulletin/bulletin.html.
Centers for Disease Control. 2013. HIV in the United States at a Glance. http://www.cdc.gov/hiv/statistics/basics/ataglance.html. Accessed on August 27, 2013.
Charlton, James I. 1998. *Nothing About Us Without Us: Disability Oppression and Empowerment*. Berkeley: University of California Press.
Chaudhuri, Nupur. 1988. "Memsahibs and Motherhood in Nineteenth-Century Colonial India." *Victorian Studies* 31(4): 517–35.
Chen, Mel. 2012. *Animacies: Biopolitics, Racial Mattering, and Queer Affect*. Durham, NC: Duke University Press.
Chik, F. and S. A. Rabbani. 2011. "Role of Epigenetics in Cancer Initiation and Progression." *Advances in Experimental Medicine and Biology* 720: 91–104.
Chilibeck, Philip D., Digby G. Sale, and Colin E. Webber. 1995. "Exercise and Bone Mineral Density." *Sports Medicine* 19(2): 103–22.
Chinnery, Patrick F. and Douglass M. Turnbull. 2001. "Epidemiology and Treatment of Mitochondrial Disorders." *American Journal of Medical Genetics* 106(1): 94–101.
Chiò, A., G. Logroscino, B. J. Traynor, J. Collins, J. C. Simeone, L. A. Goldstein, and L. A. White. 2013. "Global Epidemiology of Amyotrophic Lateral Sclerosis: A Systematic Review of the Published Literature." *Neuroepidemiology* 41(2): 118–30.
Chodorow, Nancy. 1978. The Reproduction of Mothering: Psychoanalysis and the Sociology of Gender. Berkeley: University of California Press.

Choi, Precilla. 2003. "Muscle Matters: Maintaining Visible Differences between Women and Men." *Sexualities, Evolution and Gender* 5(2): 71–81.
Chouinard, Vera. 2013. "Precarious Lives in the Global South: On Being Disabled in Guyana." *Antipode* 46(2), September.
Christakis, Dimitri A., Eric Harvey, Danielle M. Zerr, et al. 2000. "A Trade-Off Analysis of Routine Newborn Circumcision." *Journal of the Ambulatory Pediatric Association* (Supplement) 105: 246–49.
Christensen, Vivian A. and Monica J. Casper. 2000. "Hormone Mimics and Disrupted Bodies: A Social Worlds Analysis of a Scientific Controversy." *Sociological Perspectives* 43: S93–S120.
Claeys, George. 2000. "The 'Survival of the Fittest' and the Origins of Social Darwinism." *Journal of the History of Ideas* 61(2): 223–40.
Clarey, Christopher and Gina Kolata. 2009. "Gold Awarded Amid Dispute over Runner's Sex." *New York Times*, August 21.
Clark, Brett and Richard York. 2005. "Carbon Metabolism: Global Capitalism, Climate Change, and the Biospheric Rift." *Theory & Society* 34(4): 391–428.
Clarke, Adele E. 1984. "Subtle Forms of Sterilization Abuse: A Reproductive Rights Analysis." pp. 188–212 in *Test-tube Women: What Future for Motherhood?*, edited by Rita Arditti, Renate Duelli Klein, and Shelley Minden. London: Pandora Press.
Clarke, Adele E. 1995. "Modernity, Postmodernity, and Reproductive Processes, ca. 1890–1990: Or 'Mommy, where do cyborgs come from anyway?'" pp. 139–56 in *The Cyborg Handbook*, edited by Chris Hables Gray, Heidi Figueroa-Sarriera, and Steven Mentor. New York: Routledge.
Clarke, Adele E. 1998. *Disciplining Reproduction: Modernity, American Life Sciences, and the Problems of Sex*. Berkeley: University of California Press.
Clarke, Adele E. and Virginia L. Olesen. 1998. *Revisioning Women, Health and Healing: Feminist, Cultural, and Technoscience Perspectives*. New York: Routledge.
Clough, Patricia T. 2010. "The Affective Turn: Political Economy, Biomedia, and Bodies." pp. 206–25 in *The Affect Theory Reader*, edited by Melissa Gregg and Gregory J. Seigworth. Durham, NC: Duke University Press.
Cohen, I. Glenn. 2013. "Transplant Tourism: The Ethics and Regulation of International Markets for Organs." *Journal of Law, Medicine and Ethics* 41(1): FIX.
Cohen, Jon. 2006. "The New World of Global Health." *Science* 311: 162–67.
Cohen, Lawrence. 1998. *No Aging in India: Alzheimer's, the Bad Family, and Other Modern Things*. Berkeley: University of California Press.
Cohn, Amanda C. and Lee H. Harrison. 2013. "Meningococcal Vaccines: Current Issues and Future Strategies." *Drugs* 73(11): 1147–55.
Cohn, Jonathan. 2011. "The Two Year Window." *New Republic*, November 9.
Colen, Shellee. 1986. "'Like a Mother to Them': Stratified Reproduction and West Indian Childcare Workers and Employers in New York." pp. 78–102 in Faye D. Ginsburg and Rayna Rapp. eds. *Conceiving the New World Order: The Global Politics of Reproduction*. Berkeley: University of California Press.
Collier, Aine. 2007. *The Humble Little Condom: A History*. Amherst, NY: Prometheus Books.

Colon, I., D. Caro, C. J. Bourdony, and O. Rosario. 2000. "Identification of Phthalate Esters in the Serum of Young Puerto Rican Girls with Premature Breast Development." *Environmental Health Perspectives* 108(9): 895–900.

Connell, R. W. 1995. *Masculinities*. Berkeley: University of California Press.

Coole, Diana and Samantha Frost. 2010. *New Materialisms: Ontology, Agency, and Politics*. Durham, NC: Duke University Press.

Cooley, Charles Horton. 1902. *Human Nature and the Social Order*. New York: Scribner's.

Cozolino, Louis. 2006. *The Neuroscience of Human Relationships: Attachment and the Developing Social Brain*. New York: W.W. Norton & Company.

Craviari, Thierry, John M. Pettifor, Tom D. Thacher, Craig Meisner, Josiane Arnaud, and Philip R. Fischer. 2008. "Rickets: An Overview and Future Directions, with Special Reference to Bangladesh." *Journal of Health, Population and Nutrition* 26(1): 112–21.

Crawford, Bridget. 2011. "The Currency of White Women's Hair in a Down Economy." (January 25, 2011). *Women's Rights Law Reporter*. Available at SSRN: http://ssrn.com/abstract=1748103.

Creighton, Sarah and Catherine Minto. 2001. "Managing Intersex: Most Vaginal Surgery in Childhood Should be Deferred." *British Medical Journal* 323(7324): 1264–65.

Crenshaw, Kimberlé W. 1991. "Mapping the Margins: Intersectionality, Identity Politics, and Violence against Women of Color." *Stanford Law Review* 43(6): 1241–99.

Croteau, David, William Hoynes, and Stefania Milan, eds. 2012. *Media Society: Industries, Images, and Audiences*. 4th edition. London: Sage Publications.

Currah, Paisley. 2006. "Gender Pluralisms Under the Transgender Umbrella." In *Transgender Rights*, edited by Paisley Currah, Richard M. Juang, and Shannon Price Minter. Minneapolis: University of Minnesota Press.

Currah, Paisley. 2008. "Expecting Bodies: The Pregnant Man and Transgender Exclusion from the Employment Non-Discrimination Act." *Women's Studies Quarterly* 36(3&4): 330–36.

Curtis, Ted, Robert Dellar, and Leslie Esther, eds. 1999. *Mad Pride: A Celebration of Mad Culture*. London: Chipmunka Publishing.

Cutler, David and Grant Miller. 2005. "The Role of Public Health Improvements in Health Advances: The 20th Century United States." *Demography* 42(1): 1–22. http://www.economics.harvard.edu/files/faculty/13_The%20Role%20of%20Public%20Health%20Improvements%20in%20Health%20Advances.pdf.

Danaei, Goodarz, Stephen Vander Hoorn, Alan D. Lopez, Christopher J. L. Murray, and Majid Ezzati. 2005. "Causes of Cancer in the World: Comparative Risk Assessment of Nine Behavioural and Environmental Risk Factors." *The Lancet* 366(9499): 1784–93.

Danovitch, G. M. and M. E. Shapiro. 2011. "The Use of Executed Prisoners as a Source of Organ Transplants in China Must Stop." *American Journal of Transplantation* 11(3): 426–28.

Davey, Belinda. 2006. "Pain During Mammography: Possible Risk Factors and Ways to Alleviate Pain." *Radiography* 13: 229–34.

Davis, Angela Y. 1981. *Women, Race, and Class*. New York: Random House.

Davis, Dana-Ain. 2009. "The Politics of Reproduction: The Troubling Case of Nadya Suleman and Assisted Reproductive Technology." *Transforming Anthropology* 17(2): 105–16.

Davis, Devra. 2007. *The Secret History of the War on Cancer*. New York: Basic Books.

Davis, Kathy. 2007. *The Making of Our Bodies, Ourselves: How Feminism Travels Across Borders*. Durham, NC: Duke University Press.

Davis, Lennard J., ed. 2013. *The Disability Studies Reader*. 4th edition. New York: Routledge.

Davis-Floyd, Robbie and Eugenia Georges. 1996. "On Pregnancy." pp. 1014–16 in *Encyclopedia of Cultural Anthropology*. New Haven, CT: Human Relations Area Files.

Davis-Kimball, Jeannine and Mona Behan. 2003. *Warrior Women: An Archaeologist's Search for History's Hidden Heroines*. New York: Warner Books.

de Bernis, L. 2007. "Obstetric Fistula: Guiding Principles for Clinical Management and Programme Development, a New WHO Guideline." *International Journal of Gynecology and Obstetrics* 99: S117–S121.

De Lora, P., and Blanco, A. P. 2013. "Dignifying Death and the Morality of Elective Ventilation." *Journal of Medical Ethics* 39(3): 145–48.

de Wert, Guido and Christine Mummery. 2003. "Human Embryonic Stem Cells: Research, Ethics and Policy." *Human Reproduction* 18(4): 672–82.

Delmonico, Francis L., Robert Arnold, Nancy Scheper-Hughes, Laura A. Siminoff, Jeffrey Kahn, and Stuart J. Youngner. 2002. "Ethical Incentives—Not Payment—For Organ Donation." *The New England Journal of Medicine* 346(25): 2002–05.

Dementia Research Group, ETATS-UNIS. 2004. "Care Arrangements for People with Dementia in Developing Countries." *International Journal of Geriatric Psychiatry* 19(2): 170–77.

Devault, Marjorie L. 1994. *Feeding the Family: The Social Organization of Caring as Gendered Work*. Chicago: University of Chicago Press.

Diamond, Lisa. 2009. *Sexual Fluidity: Understanding Women's Love and Desire*. Cambridge, MA: Harvard University Press.

Diamond, Tim. 1995. *Making Gray Gold: Narratives of Nursing Home Care*. Chicago: University of Chicago Press.

Dihn, Quan and Rodney Sinclair. 2007. "Female Pattern Hair Loss: Current Treatment Concepts." *Clinical Interventions in Aging* 2(2): 189–99.

Doidge, Norman. 2007. *The Brain That Changes Itself: Stories of Personal Triumph from the Frontiers of Brain Science*. New York: Penguin Books.

Douglas, Mary. 2002 [1966]. *Purity and Danger: An Analysis of Concepts of Pollution and Taboo*. New York: Routledge.

Douglas, Susan and Meredith Michaels. 2005. *The Mommy Myth: The Idealization of Motherhood and How It Has Undermined All Women*. New York: Free Press.

Drabiak-Syed, Katherine. 2010. "Lessons from Havasupai Tribe vs. Arizona State University Board of Regents: Recognizing Group, Cultural, and Dignitary Harms as Legitimate Risks Warranting Integration into Research Practice." *Journal of Health and Biomedical Law* 1: 175–225.

Draganski, Bogdan, Christian Gaser, Volker Busch, Gerhard Schuierer, Ulrich Boghdan, and Arne May. 2004. "Neuroplasticity: Changes in Grey Matter Induced by Training." *Nature* 427: 311–12.

Dreger, Alice. 2009. "Where's the Rulebook for Sex Verification?" *New York Times*, August 22, D1, 3.

Drew, Shirley, Melanie Mills and Bob M. Gassaway. 2007. *Dirty Work. The Social Construction of Taint*. Waco, TX: Baylor University Press.

Drife, J. 2002. "The Start of Life: A History of Obstetrics." *Postgraduate Medical Journal* 78: 311–15.

Drummond, Katie. 2012. "What Role Did Accused Soldier's Brain Injury Play in Afghan Massacre?" *Wired*, March 13. http://www.wired.com/dangerroom/2012/03/afghan-massacre-tbi/.

Dubois, W. E. B. 1903. *The Souls of Black Folk*. Chicago: AC McClurg and Company.

Dubriwny, Tasha N. 2009. "Constructing Breast Cancer in the News: Betty Ford and the Evolution of the Breast Cancer Patient." *Journal of Communication Inquiry* 33(2): 104–25.

Duden, Barbara. 1993. *Disembodying Women: Perspectives on Pregnancy and the Unborn*. Cambridge, MA: Harvard University Press.

Duggan, Lisa and Nan D. Hunter. 1995. *Sex Wars: Sexual Dissent and Political Culture*. New York: Routledge.

Dumit, Joseph. 2012. *Drugs for Life: How Pharmaceutical Companies Define Our Health*. Durham, NC: Duke University Press.

Duneier, Mitchell. 1994. *Slim's Table: Race, Respectability, and Masculinity*. Chicago: University of Chicago Press.

Duster, Troy. 2003 [1990]. *Backdoor to Eugenics*. 2nd edition. New York: Routledge.

Dworkin, Andrea. 1981. *Pornography: Men Possessing Women*. London: The Women's Press.

Dworkin, Sharon and Faye Wachs. 2009. *Body Panic: Gender, Health and the Selling of Fitness*. New York: New York University Press.

Elias, Norbert. 1994. *The Civilizing Process: The History of Manners and State Formation*. Cambridge: Polity Press.

Ellis, Juniper. 2008. *Tattooing the World: Pacific Designs in Print and Skin*. New York: Columbia University Press.

Erickson, E. J., S. W. Merrell, J. R. Saffle, and J. J. Sullivan. 1991. "Differences in Mortality from Thermal Injury Between Pediatric and Adult Patients." *Journal of Pediatric Surgery* 26(7): 821–25.

Evans, R. W. 2008. "Ethnocentrism Is an Unacceptable Rationale for Health Care Policy: A Critique of Transplant Tourism Position Statements." *American Journal of Transplantation* 8(6): 1089–95.

Fagan, Honor. 2002. "Grounding Waste." National Institute for Regional and Spatial Analysis, Working Paper Series, No. 1, 2002: NUI Maynooth.

Fahs, Breanne. 2012. "Breaking Body Hair Boundaries: Classroom Exercises for Challenging Social Constructions of the Body and Sexuality." *Feminism and Psychology* 22(4): 482–506.

Fainaru-Wada, Mark and Steve Fainaru. 2013. *League of Denial: The NFL, Concussions, and the Battle for Truth*. New York: Crown Archetype.

Fallon, Patricia, Melanie A. Katzman, and Susan C. Wooley, eds. 1996. *Feminist Perspectives on Eating Disorders*. 1st edition. New York: Guilford Press.

Falomir-Pichastor, Juan and Peter Hegarty. 2013. "Maintaining Distinctions under Threat: Heterosexual Men Endorse the Biological Theory of Sexuality when Equality Is the Norm." *British Journal of Social Psychology* DOI: 10.1111/bjso.12051

Farmer, Paul. 2005. *Pathologies of Power: Health, Human Rights, and the New War on the Poor*. Berkeley: University of California Press.

Fausto-Sterling, Anne. 2005. "The Bare Bones of Sex: Part I—Sex and Gender." *Signs* 30(2): 1491–1527.

Fausto-Sterling, Anne. 2008. "The Bare Bones of Race." *Social Studies of Science* 38(5): 657–94.

Featherstone, Mike. 1982. "The Body in Consumer Culture." *Theory, Culture and Society* 1(2): 18–33.

Feder, Ellen K. and Katrina Karkazis. 2008. "What's In a Name? The Controversy Over 'Disorders of Sex Development.'" *Hastings Center Report* 38(5): 33–36.

Feilden, Tom. 2012. "A Golden Age of Discovery in Neuroscience." BBC News, September 5.

Fenton, Andrew and Tim Krahn. 2007. "Autism, Neurodiversity, and Equality Beyond the 'Normal.'" *Journal of Ethics in Mental Health* 2(2): 1–6.

Fields, Jill. 2007. *An Intimate Affair: Women, Lingerie, and Sexuality*. Berkeley: University of California Press.

Figert, Anne. 1996. *Women and the Ownership of PMS: The Structuring of a Psychiatric Disorder*. Piscataway, NJ:Transaction Publishers.

Firestone, Shulamith. 1970. *The Dialectic of Sex: The Case for Feminist Revolution*. New York: William Morrow & Company.

Fishman, Jennifer R. 2004. "Manufacturing Desire: The Commodification of Female Sexual Dysfunction." *Social Studies of Science* 34(2): 187–218.

Fortun, Michael A. 2008. *Promising Genomics: Iceland and deCode Genetics in a World of Speculation*. Berkeley: University of California Press.

Foster, John Bellamy (1999). "Marx's Theory of Metabolic Rift: Classical Foundations for Environmental Sociology." *The American Journal of Sociology* 105(2): 381.

Foucault, Michel. 1976 [1990]. *History of Sexuality, Volume 1*. New York: Vintage Books.

Fountain, Henry. 2012. "Body Builders: Making Artificial Organs from Cells." *New York Times*, December 19.

Frank, Arthur W. 1991. *At the Will of the Body: Reflections on Illness*. New York: Houghton Mifflin.

Franklin, Sarah. 2007. *Dolly Mixtures: The Remaking of Genealogy*. Durham, NC: Duke University Press.

Franklin, Sarah and Celia Roberts. 2006. *Born and Made: An Ethnography of Preimplantation Genetic Diagnosis*. Princeton: Princeton University Press.
Freidberg, Susanne. 2010. *Fresh: A Perishable History*. Cambridge, MA: Belknap Press of Harvard University Press.
Fulambarker, A., Farooki, B., Kheir, F., Copur, A. S., Srinivasan, L., and Schultz, S. 2012. "Effect of Yoga in Chronic Obstructive Pulmonary Disease." *American Journal of Therapeutics* 19(2): 96–100.
Fussell, Samuel Wilson. 1991. *Muscle: Confessions of an Unlikely Bodybuilder.* New York: Avon Books.
Gakidou, Emmanuela, Stella Nordhagen, and Ziad Obermeyer. 2008. "Coverage of Cervical Cancer Screening in 57 Countries: Low Average Levels and Large Inequalities." *PLoS Medicine* 5(6): e132.
Galarneau, Charlene. 2010. "Blood Donation, Deferral, and Discrimination: FDA Donor Deferral Policy for Men Who Have Sex With Men." *The American Journal of Bioethics* 10(2): 29–39.
Gallagher, James. 2011. "Gay Men Blood Donor Ban to be Lifted." *BBC News*, September 8.
Gallucci, Stefania and Polly Matzinger. 2001. "Danger Signals: SOS to the Immune System." *Current Opinion in Immunology* 13(1): 114–19.
Gandy, M., 2004. "Rethinking Urban Metabolism: Water, Space and the Modern City." *City* 8(3): 363–79.
Gaskin, Ina May. 2011. *Birth Matters: A Midwife's Manifesta*. New York: Seven Stories Press.
Lavie, Smadar and Ted Sweden burg, editors. 1996. *Displacement, Diaspara, and Geographies of Identity*. Durham & London: Duke University Press.
Giddens, Anthony. 1991. *Modernity and Self-Identity: Self and Society in the Late.* Modern Age. Palo Alto: Stanford University Press.
Gilmore, Ruth Wilson. 2007. *Golden Gulag: Prisons, Surplus, Crisis, and Opposition in Globalizing California*. Berkeley: University of California Press.
Ginley, Barry. 2013. "Museums: A Whole New World for Visually Impaired People." *Disability Studies Quarterly* 33: 3.
Glaser, Gabrielle. 2008. "'Mad Pride' Fights a Stigma." *New York Times*, May 11.
Glenn, Evelyn Nakano. 1993. "Social Constructions of Mothering: A Thematic Overview." pp. 1–29 in *Mothering: Ideology, Experience, and Agency*, edited by Evelyn Nakano Glenn, Grace Chang, and Linda Rennie Forcey. New York: Routledge.
Glenn, Evelyn Nakano. 2008. "Yearning for Lightness: Transnational Circuits in the Marketing and Consumption of Skin Lighteners." *Gender and Society* 22(3): 281–302.
Gloeckner, Mary Reid. 1984. "Perceptions of Sexual Attractiveness Following Ostomy Surgery." *Research in Nursing and Health* 7: 87–92.
Glover, Karen S. 2009. *Racial Profiling: Research, Racism, and Resistance*. New York: Rowman & Littlefield Publishers.
Goffman, Erving. 1959. *The Presentation of Self in Everyday Life*. New York: Doubleday.

Goffman, Erving. 1961. *Asylums: Essays on the Social Situations of Mental Patients and Other Inmates*. Aldine: Chicago.

Goffman, Erving. 1963. *Stigma: Notes on the Management of Spoiled Identity*. Englewood Cliffs, NJ: Prentice Hall.

Goffman, Erving. 1967. *Interaction Ritual: Essays on Face-to-Face Behavior*. New York: Anchor Books.

Golden, Janet. 2001. *Social History of Wet Nursing in America: From Breast to Bottle*. Columbus, OH: Ohio State University Press.

Goldstein, M. 1994. "Decade of the Brain: An Agenda for the Nineties." *Western Journal of Medicine* 161(3): 239–41.

Goodman, Steve. 2009. *Sonic Warfare: Sound, Affect, and the Ecology of Fear*. Cambridge, MA: MIT Press.

Gordon-Dseagu, Vanessa. 2006. "Cancer and Health Inequalities: An Introduction to Current Evidence." Cancer Research UK. http://www.cancerresearchuk.org/prod_consump/groups/cr_common/@nre/@pol/documents/generalcontent/crukmig_1000ast-3344.pdf.

Gould, Stephen Jay. 1981. *The Mismeasure of Man*. New York: W.W. Norton & Company.

Grafton, Anthony, Glenn W. Most, and Salvatore Settis, eds. 2010. *The Classical Tradition*. Cambridge, MA: Harvard University Press.

Grandin, Temple. 2010. *Animals Make Us Human: Creating the Best Life for Animals*. New York: Mariner Books.

Grant, Jaime M., Lisa A. Mottet, Justin Tanis, Jack Harrison, Jody L. Herman, and Mara Keisling. 2011. "Injustice at Every Turn: A Report of the National Transgender Discrimination Survey." Washington: National Center for Transgender Equality and National Gay and Lesbian Task Force. www.thetaskforce.org/downloads/reports/ntds_full.pdf

Graper, C., M. Milne and M. R. Stevens. 1996. "The Traumatic Saddle Nose Deformity: Etiology and Treatment." *Journal of Craniomaxillofacial Trauma* 2(1): 50.

Gravlee, Clarence. 2009. "How Race Becomes Biology: Embodiment of Social Inequality." *American Journal of Physical Anthropology* 139: 47–57.

Greenblatt, Samuel H. 1995. "Phrenology in the Science and Culture of the 19th Century." *Neurosurgery* 37(4): 790–805.

Greenhalgh, Susan. 2008. *Just One Child: Science and Policy in Deng's China*. Berkeley: University of California Press.

Greil, Arthur L., Julia McQuillan, Karina M. Shreffler, Katherine M. Johnson, and Kathleen S. Slauson-Blevins. 2011. "Race-Ethnicity and Medical Services for Infertility: Stratified Reproduction in a Population-Based Sample of U.S. Women." *Journal of Health and Social Behavior* 52: 493–507. To access, http://digitalcommons.unl.edu/cgi/viewcontent.cgi?article=1167&context=sociology facpub.

Griffiths, R. Wayne, J. David Bigelow, and James Loewen. 2010. "Foreskin Restoration 1980–2008." *Genital Anatomy* 189–98.

Grob, Rachel. 2011. *Testing Baby: The Transformation of Newborn Screening, Parenting, and Policymaking*, New Brunswick, NJ: Rutgers University Press.

Grosz, Elizabeth. 1994. *Volatile Bodies: Toward a Corporeal Feminism*. Bloomington: Indiana University Press.

Grosz, Elizabeth. 1995. *Space, Time, and Perversion: Essays on the Politics of Bodies*. New York: Routledge.

Guice, Jon. 1998. "Controversy and the State: Lord ARPA and Intelligent Computing." *Social Studies of Science* 28(1): 103–38.

Guthman, Julie. 2011. *Weighing In: Obesity, Food Justice, and the Limits of Capitalism*. Berkeley: University of California Press.

Guyer, Sara. 2009. "Rwanda's Bones." *boundary 2* 36(2): 155–75.

Hadad, Lisa B. and Nawal M. Nour. 2009. "Unsafe Abortion: Unnecessary Maternal Mortality." *Review of Obstetrics and Gynecology* 2(2). To access: http://www.ncbi.nlm.nih.gov/pmc/articles/PMC2709326/.

Hamer, Dean. 1994. *The Science of Desire: The Search for the Gay Gene and the Biology of Behavior*. New York: Simon and Schuster.

Hamilton, Brady E., Joyce A. Martin, and Stephanie J. Ventura. 2011. "Births: Preliminary Data for 2010." *National Vital Statistics Reports* 60(2): 1–25. To access: http://www.cdc.gov/nchs/data/nvsr/nvsr60/nvsr60_02.pdf.

Hamilton, David. 2012. *A History of Organ Transplantation: Ancient Legends to Modern Practice*. Pittsburgh: University of Pittsburgh Press.

Hammonds, Evelyn M. and Rebecca M. Herzig, eds. 2009. *The Nature of Difference: Sciences of Race in the United States from Jefferson to Genomics*. Boston: The MIT Press.

Haraway, Donna. 1988. "Situated Knowledges: The Science Question in Feminism and the Privilege of Partial Perspective." *Feminist Studies* 14(3): 575–99.

Haraway, Donna. 1989. *Primate Visions: Gender, Race, and Nature in the World of Modern Science*. New York: Routledge.

Haraway, Donna. 1991. *Simians, Cyborgs and Women: The Reinvention of Nature*. New York: Routledge.

Haraway, Donna. 1997. *Modest_Witness@Second_Millenium.FemaleMan_Meets_Onco Mouse: Feminism and Technoscience*. New York: Routledge.

Harmon, Amy. 2010. "Indian Tribe Wins Fight to Limit Research of Its DNA." *New York Times*, April 21.

Harvey, David. 2007. *A Brief History of Neoliberalism*. New York: Oxford University Press.

Hawthorne, Susan. 2007. "Land, Bodies, and Knowledge: Biocolonialism of Plants, Indigenous Peoples, Women, and People with Disabilities." *Signs* 32(2): 314–23.

Hayles, N. Katharine. 1999. *How We Became Posthuman: Virtual Bodies in Cybernetics, Literature and Informatics*. Chicago: University of Chicago Press.

Hedgecoe, Adam. 2004. *The Politics of Personalized Medicine: Pharmacogenetics in the Clinic*. Cambridge: Cambridge University Press.

Henderson, B. E., B. Benton, M. Cosgrove, J. Baptista, J. Aldrich, D. Townsend, W. Hart, and T. M. Mack. 1976. "Urogenital Tract Abnormalities in Sons of Women Treated with Diethylstilbestrol." *Pediatrics* 58(4): 505–507.

Henneh, William, Pippa Thompson, Leena Peltonen, and David Porteous. 2006. "Genes and Schizophrenia: Beyond Schizophrenia: The Role of *DISC1* in Major Mental Illness." *Schizophrenia Bulletin* 32(3): 409–16.

Henze, K., W. Martin, and William Martin. 2003. "Evolutionary Biology: Essence of Mitochondria." *Nature* 426(6963): 127–28.

Hernandez, Kelly Lytle. 2010. *Migra! A History of the U.S. Border Patrol.* Berkeley: University of California Press.

Hesketh, Therese, Li Lu, and Zhu Wei Xing. 2005. "The Effect of China's One-Child Family Policy After 25 Years." *New England Journal of Medicine* 353: 1171–76.

Hess, Amanda. 2013. "Put the Pill on Drugstore Shelves. Pregnancy is More Dangerous than Birth Control." *Slate*, April 23. To access: http://www.slate.com/blogs/xx_factor/2013/04/23/over_the_counter_birth_control_pills_oral_contraceptives_are_safer_than.html.

Heyes, Cressida J. and Meredith Jones, eds. 2009. *Cosmetic Surgery: A Feminist Primer.* Surrey: Ashgate.

Hoeyer, Klaus. 2008. "The Ethics of Research Biobanking: A Critical Review of the Literature." *Biotechnology and Genetic Engineering Reviews* 25: 429–52.

Holdcroft, Anita. 2007. "Gender Bias in Research: How Does It Affect Evidence Based Medicine?" *Journal of the Royal Society of Medicine* 100(1): 2–3.

Hook, G. Raumati. 2009. "'Warrior Genes' and the Disease of Being Maori." *MAI Review* 2: 1–22.

hooks, bell. 1998. "Straightening My Hair." *Z Magazine.*

Hoyez, Anne-Cecile. 2007. "The 'world of yoga': The Production and Reproduction of Therapeutic Landscapes." *Social Science and Medicine* 65(1): 112–24.

Hughes, Everett C. 1951. "Work and the Self," in J. H. Roher and M. Sherif (eds.), Social Psychology at the Crossroads, New York: Harper, pp. 313–23 in *Social Psychology at the*

Hughes, Everett C. 1971. *The Sociological Eye: Selected Papers.* Chicago: Aldine-Atherton.

Hughes, J. 1998. "Xenografting: Ethical Issues." *Journal of Medical Ethics* 24: 18–24.

Hunter, Margaret. 2002. "If You're Light You're Alright" Light Skin Color as Social Capital for Women of Color." *Gender and Society* 16(2): 175–93.

Huxley, Aldous. 1932. *Brave New World.* London: Chatto & Windus.

Hvistendahl, Mara. 2012. "Cesarean Nation: The Cautionary Tale of How China Came to Have the World's Highest C-Section Rate." *Slate*, January 3. To access: http://www.slate.com/articles/double_x/doublex/2012/01/cesarean_nation_why_do_nearly_half_of_chinese_women_deliver_babies_via_c_section_.html.

Hyder, Adnan A., Colleen A. Wunderlich, Prasanthi Puvanachandra, G. Gururaj, and Olive C. Kobusingye. 2007. "The Impact of Traumatic Brain Injuries: A Global Perspective." *NeuroRehabilitation* 22(5): 341–53.

Hyman, Steven E. and Robert C. Malenka. 2001. "Addiction and the Brain: The Neurobiology of Compulsion and Its Persistence." *Nature Reviews Neuroscience* 2: 695–703.

ICSCN. 2008. "Global Regulation of Human Embryonic Stem Cell Research and Oocyte Donation." International Consortium of Stem Cell Networks,

http://icscn.files.wordpress.com/2008/09/global-regulation-hesc-research-oocyte-donation-sep-08.pdf.
Immergut, Matthew. 2010. "Manscaping: The Tangle of Nature, Culture and Male Body Hair." in *The Body Reader: Essential Social and Cultural Readings*, edited by Lisa Jean Moore and Mary Kosut. New York: New York University Press.
Imray, Gerald. 2013. "Reeva Steenkamp Murder: Oscar Pistorius to Face More Charges." *Christian Science Monitor*, October 29.
Inda, Jonathan Xavier, ed. 2005. *Anthropologies of Modernity: Foucault, Governmentality, and Life Politics*. Malden, MA: Wiley-Blackwell.
Inda, Jonathan Xavier. 2011. "Materializing Hope: Racial Pharmaceuticals, Suffering Bodies, and Biological Citizenship." In Monica J. Casper and Paisley Currah, eds. *Corpus: An Interdisciplinary Reader on Bodies and Knowledge*. New York: Palgrave Macmillan.
Institute of Medicine Committee on Understanding Premature Birth and Assuring Healthy Outcomes. 2007. *Preterm Birth: Causes, Consequences, and Prevention*. Edited by R. E. Behrman and A. S. Butler. Washington, DC: National Academies Press. To access: http://www.ncbi.nlm.nih.gov/pubmed/20669423.
Iredale, Rachel, Kate Brain, Buddug Williams, Elizabeth France, and Jonathon Gray. 2005. "The Experiences of Men with Breast Cancer in the United Kingdom." *European Journal of Cancer* 42(3): 334–41.
Isa, Masako and Eric Mark Kramer. 2003. "Adopting the Caucasian 'Look': Reorganizing the Minority face." In *The Emerging Monoculture: Assimilation and the "Model Minority"* edited by Mark Kramer. Westport: Praeger Press: 41–74.
ISAPS. 2012. "ISAPS International Survey on Aesthetic/Cosmetic Procedures Performed in 2011." To access: http://www.isaps.org/files/html-contents/Downloads/ISAPS%20Results%20-%20Procedures%20in%202011.pdf.
ISSCR. 2006. "Guidelines for the Conduct of Human Embryonic Stem Cell Research." International Society for Stem Cell Research, December 1. http://www.isscr.org/docs/default-source/hesc-guidelines/isscrhescguidelines2006.pdf.
Jacobson, Nora. 1999. *Cleavage: Technology, Controversy, and the Ironies of the Man-Made Breast*. New Brunswick, NJ: Rutgers University Press.
Jagose, Annamarie. 1996. *Queer Theory: An Introduction*. New York: New York University Press.
Jahoda, Gustav. 2007. "Toward Scientific Racism." In *Race and Racialization*. edited by Tania Das Gupta. Toronto: Canadian Scholars Press. pp. 24–30.
Jarrett, Christian. 2012. "Why the Left-Brain Right-Brain Myth Will Probably Never Die." *Psychology Today*, June 27.
Jasny, Barbara R., Katrina L. Kellner, and Elizabeth Pennisi. 2008. "From Genes to Social Behavior: Introduction to Special Issue." *Science* 322(5903): 891.
Jauhar, Sandeep. 2004. "The Artificial Heart." *New England Journal of Medicine* 350: 542–44.
Jemal, Ahmedin, Freddie Bray, Melissa M. Center, Jacques Ferlay, Elizabeth Ward, and David Forman. 2011. "Global Cancer Statistics." *CA: A Cancer Journal for Clinicians* 61(2): 69–90.

Jian, Ma. 2013. "China's Brutal One-Child Policy." *New York Times*, May 21. To access: http://www.nytimes.com/2013/05/22/opinion/chinas-brutal-one-child-policy.html?_r=0.

Joffe, Carole E. 1996. *Doctors of Conscience: The Struggle to Provide Abortion Before and After Roe v. Wade*. Boston: Beacon Press.

Jolie, Angelina. 2013. "My Medical Choice." *New York Times*, May 14.

Jones-Rogers, Stephanie E. 2011. "Black Milk: Maternal Bodies, Wet Nursing, and the Value of Black Women's Invisible Labor in the Antebellum Slave Market." Berkshire Conference of Women Historians, December 6. http://scholarworks.umass.edu/berksconference/Workshops/179/1/.

Jones, Lisa. 1997. *Bulletproof Diva: Tales of Race, Sex, and Hair*. New York: Anchor.

Jonsson, Gudrun. 2013. *Gut Reaction: A Day-by-Day Programme for Choosing and Combining Foods for Better Health and Easy Weight Loss*. New York: Random House.

Jordan, Brigitte and Robbie Davis-Floyd. 1992. *Birth in Four Cultures: A Crosscultural Investigation of Childbirth in Yucatan, Holland, Sweden, and the United States*. 4th edition. Long Grove, IL: Waveland Press.

Kapsalis, Terri. 1997. *Public Privates: Performing Gynecology from Both Ends of the Spectrum*. Durham, NC: Duke University Press.

Karkazis, Katrina. 2008. *Fixing Sex: Intersex, Medical Authority, and Lived Experience*. Durham, NC: Duke University Press.

Kessel, Anna. 2009. "Gold Medal Athlete Caster Semenya Told to Prove She is a Woman." *The Guardian*, 19 August. http://www.guardian.co.uk/sport/2009/aug/19/castersemenya-gender-verification-test.

Kessler, David A. 1992. "The Basis of the FDA's Decision on Breast Implants." *New England Journal of Medicine* 326: 1713–15.

Khaleeli, Homa. 2012. "The Hair Trade's Dirty Secret." *The Guardian*, 28 October 2012.http://www.guardian.co.uk/lifeandstyle/2012/oct/28/hair-extension-global-trade-secrets.

Kimmel, Michael. 1996. *Changing Men: New Directions in the Study of Men and Masculinity*. Newbury Park, CA: Sage Publications.

King-O'Riain, Rebecca Chiyoko. 2006. *Pure Beauty: Judging Race in Japanese American Beauty Pageants*. Minneapolis: University of Minnesota Press.

King, Samantha. 2008. *Pink Ribbons, Inc.: Breast Cancer and the Politics of Philanthropy*. Minneapolis: University of Minnesota Press.

Kirby, David. 2011. *Animal Factory: The Looming Threat of Pig, Dairy, and Poultry Farms to Humans and the Environment*. New York: St. Martin's Griffin.

Klawiter, Maren. 2008. *The Biopolitics of Breast Cancer: Changing Cultures of Disease and Activism*. Minneapolis: University of Minnesota Press.

Kleese, Christian. 1999. "Modern Primitivism: Non-mainstream Body Modification and Racialized Representation." *Body and Society* 5 (2/3): 15–38.

Klein, Alan M. 1993. *Little Big Men: Bodybuilding Subculture and Gender Construction*. Albany, NY: SUNY Press.

Kleinman, Arthur. 1987. "Anthropology and Psychiatry: The Role of Culture in Cross-Cultural Research on Illness." *The British Journal of Psychiatry* 151: 447–54.

Kligman, Gail. 1995. "Political Demography: The Banning of Abortion in Ceausescu's Romania." pp. 234–55 in *Conceiving the New World Order: The Global Politics of Reproduction*, edited by Faye D. Ginsburg and Rayna Rapp. Berkeley: University of California Press.

Kligman, Gail. 1998. *The Politics of Duplicity: Controlling Reproduction in Ceausescu's Romania*. Berkeley: University of California Press.

Kluchin, Rebecca M. 2011. *Fit to Be Tied: Sterilization and Reproductive Rights in America, 1950–1980*. New Brunswick, NJ: Rutgers University Press.

Kluger, J., Gorman, C., and Park, A. 2004. "Why We EAT." *Time* 163(23): 72–76.

Knight, Chris. 1991. *Blood Relations: Menstruation and the Origins of Culture*. New Haven, CT: Yale University Press.

Knopf-Newman, Marcy. 2004. *Beyond Slash, Burn, and Poison: Transforming Breast Cancer Stories into Action*. New Brunswick, NJ: Rutgers University Press.

Kolassa, Iris-Tatjana and Thomas Elbert. 2007. "Structural and Functional Neuroplasticity in Relation to Traumatic Stress." *Current Directions in Psychological Science* 16(6): 321–25.

Kosut, Mary. 2006. "An Ironic Fad: The Commodificiation and Consumption of Tattoos." *The Journal of Popular Culture* 39(6): 1035–48.

Kristal, Mark B., Jean M. DiPirro, and Alexis C. Thompson. 2012. "Placentophagia in Humans and Nonhuman Mammals: Causes and Consequences." *Ecology of Food and Nutrition* 51: 177–97.

Kristeva, Julia. 1992. *Powers of Horror: An Essay on Abjection*. New York: Columbia University Press.

Kushner, T. and R. Belliotti. 1985. "Baby Fae: A Beastly Business." *Journal of Medical Ethics* 11: 178–83.

Lancet Commissions. 2009. "Managing the Health Effects of Climate Change." *The Lancet* 373: 1693–1733.

Landecker, Hannah. 2000. "Immortality, In Vitro: A History of the HeLa Cell Line." In Paul Brodwin, ed., *Biotechnology and Culture: Bodies, Anxieties, Ethics*. Bloomington: Indiana University Press.

Langrish, Jeremy et al. 2009. "Beneficial Cardiovascular Effects of Reducing Exposure to Particulate Air Pollution with a Simple Facemask." *Particle and Fibre Toxicology* 6(8). accessed http://www.biomedcentral.com/content/pdf/1743-8977-6-8.pdf.

Lanteigne, Kat. 2013. "Canada's Tainted Blood Crisis Isn't Over." *Huffington Post*, September 24.

Laqueur, Thomas W. 1990. *Making Sex: Body and Gender from the Greeks to Freud*. Cambridge, MA: Harvard University Press.

Laumann, Edward O., Christopher M. Masi, and Ezra W. Zuckerman. 1997. "Circumcision in the United States." *Journal of the American Medical Association* 277: 1052–57.

Laumann, E. O., A. Paik and R. C. Rosen. 1999. "Sexual Dysfunction in the United States: Prevalence and Predictors." *JAMA* 281: 537–544.

Lawn, Joy E., Michael G. Gravett, Toni M. Nunes, Craig E. Rubens, Cynthia Stanton, and the GAPPS Review Group. 2010. "Global Report on Preterm Birth and

Stillbirth (1 of 7): Definitions, Description of the Burden and Opportunities to Improve Data." *BMC Pregnancy and Childbirth* 10(Suppl 1): S1. To access: http://www.biomedcentral.com/1471-2393/10/S1/S1/.

Layne, Linda. 2002. *Motherhood Lost: A Feminist Account of Pregnancy Loss in America*. New York: Routledge.

Lemle, Marina. 2005. "Circumcised Men Less Likely to Get HIV, Says Study." *Science Development Network*, 29 July.

Leo, Russ. 2011. "An Archive for Affect Theory." *Reviews in Cultural Theory* 2.2, August 1.

Lerner, Barron H. 2003. *The Breast Cancer Wars: Hope, Fear, and the Pursuit of a Cure in Twentieth-Century America*. New York: Oxford University Press.

Lerner, Barron. 2009. *When Illness Goes Public: Celebrity Patients and How We Look at Medicine*. New York: Johns Hopkins University Press.

Levine, Deena R. and Mara B. Adelman. 1992. *Beyond Language: Cross Cultural Communication*. 2nd edition. Upper Saddle River, NJ: Pearson Education ESL.

Lewin, Ellen and Virginia Olesen, eds. 1985. *Women, Health, and Healing: Toward a New Perspective*. New York: Routledge.

Leys, Barbara. 2009. *From Pink to Green: Disease Prevention and the Environmental Breast Cancer Movement*. New Brunswick, NJ: Rutgers University Press.

Li, Eric, Hyun Jeong Min, Russell Beck, Junko Kimura, Shalini Bahl. 2008. "Skin Lightening and Beauty in Four Asian Cultures." *Advances in Consumer Research* 35: 444–49.

Li, San Z. and Anil K. Jain, eds. 2011. *Handbook of Face Recognition*. 2nd edition. London: Springer.

Liao, Lih-Mei, Neda Taghinejadi and Sarah Creighton. 2012. "An Analysis of the Context and Clinical Implications of Online Advertisements for Female Genital Cosmetic Surgery." *British Medical Journal* 2: 1–5.

Lindvall, Olle and Insoo Hyun. 2009. "Medical Innovation Versus Stem Cell Tourism." *Science* 324(5935): 1664–65.

Link, C. L., and McKinlay, J. B. (2009). "Disparities in the Prevalence of Diabetes: Is It Race/Ethnicity or Socioeconomic Status? Results from the Boston Area Community Health (BACH) survey." *Ethnicity & Disease* 19(3): 288–92.

Lippman, Abby. 1994. "The Genetic Construction of Prenatal Testing: Choice, Consent, or Conformity for Women?" In Rothenberg, Karen and Elizabeth Thomson, eds. *Women and Prenatal Testing: Facing the Challenges of Genetic Technology*. Columbus: Ohio State University, 9–34.

Liu, Shiliang, Robert M. Liston, K. S. Joseph, Maureen Heaman, Reg Suave, and Michael S. Kramer for the Maternal Health Study Group of the Canadian Perinatal Surveillance System. 2007. "Maternal Mortality and Severe Morbidity Associated with Low-Risk Planned Cesarean Delivery versus Planned Vaginal Delivery at Term." *Canadian Medical Association Journal* 176(4): 455–460.

Lock, Margaret 2001. *Twice Dead: Organ Transplants and the Reinvention of Death*. Berkeley: University of California Press.

Loe, Meika. 2004. *The Rise of Viagra: How The Little Blue Pill Changed Sex in America*. New York: New York University Press.

Lorang, Melissa, Bryan Callahan, Kevin Cummins, Suraj Achar and Sandra Brown. 2011. "Anabolic Androgenic Steroid Use in Teens: Prevalence, Demographics, and Perception of Effects." *Journal of Child and Adolescent Substance Abuse* 20: 358–69.
Lorber, Judith and Lisa Jean Moore. 2011. *Gendered Bodies: Feminist Perspectives.* Oxford: Oxford University Press.
Lorde, Audre. 1980. *The Cancer Journals.* Argyle, NY: Spinsters.
Lupton, Deborah. 1996. *Food, the Body and the Self.* London: Sage Publications.
Lupton, Deborah. 2013. *Fat.* New York: Routledge.
Lutz, Catherine A. and Jane L. Collins. 1993. *Reading National Geographic.* Chicago: University of Chicago Press.
Lyotard, Jean-Francois. 1984. *The Postmodern Condition: A Report on Knowledge.* Minneapolis: University of Minnesota Press.
Maasen, Sabine, E. Mendelsohn, and P. Weingart, eds. 2001. *Biology as Society, Society as Biology: Metaphors.* Sociology of the Sciences Yearbook.
Macchiarini, Paolo et al. 2008. "Clinical Transplantation of a Tissue-Engineered Airway." *The Lancet* 372: 2023–30.
MacDorman, Marian F., Eugene Declercq, Fay Menacker, and Michael H. Malloy. 2006. "Infant and Neonatal Mortality for Primary Cesarean and Vaginal Births to Women with 'No Indicated Risk,' United States, 1998–2001 Birth Cohorts." *Birth* 33(3): 175–82.
Maclean's. 2009. "Thailand: New Washroom Just for 'Ladyboys.'" 122(21): 42.
Mahmood, Khawaja Tahir, Anam Khan, and Nudrat Rafique. 2011. "Factors Contributing and Leading to Miscarriages." *Journal of Pharmaceutical Science and Technology* 3(1): 470–76.
Makino, Maria, Koji Tsuboi, and Lorraine Dennerstein. 2004. "Prevalence of Eating Disorders: A Comparison of Western and Non-Western Countries." *Medscape* 6(3): 49.
Mamo, Laura and Jennifer Ruth Fosket. 2009. "Scripting the Body: Pharmaceuticals and the (Re)Making of Menstruation." *Signs* 34(4): 925–949.
Mamo, Laura. 2007. *Queering Reproduction: Achieving Pregnancy in the Age of Technoscience.* Durham, NC: Duke University Press.
Mandelblatt, Jeanne S., Vanessa B. Sheppard, and Alfred I. Neugut. 2013. "Black-White Differences in Breast Cancer Outcomes Among Older Medicare Beneficiaries: Does Systemic Treatment Matter?" *Journal of the American Medical Association* 310(4): 376–77.
Manderson, Lenore. 2005. "Boundary Breaches: The Body, Sex and Sexuality after Stoma Surgery." *Social Science and Medicine* 61(2): 405–15.
Maranto, Gina. 2011. "Stratified Reproduction." [Review of France Winddance Twine's Outsourcing the Womb: Race, Class, and Gestational Surrogacy in a Global Market] *Biopolitical Times*, August 3. To access: http://www.biopolitical times.org/article.php?id=5806.
Margulis, Jennifer. 2013. *The Business of Baby: What Doctors Don't Tell You, What Corporations Try to Sell You, and How to Put Your Pregnancy, Childbirth, and Baby Before Their Bottom Line.* New York: Scribner's.

Marietta, Cynthia S. 2010. "Ingredient in Anti-Aging Cosmetic Re-Ignites Abortion Debate and Raises Questions about Cosmetic Labeling." Health Law Perspectives, January 1. To access: http://www.law.uh.edu/healthlaw/perspectives/2010/(CM)%20Cosmetic.pdf.

Markens, Susan. 2007. *Surrogate Motherhood and the Politics of Reproduction*. Berkeley: University of California Press.

Markoff, John. 2013. "Obama Seeking to Boost Study of Human Brain." *New York Times*, February 17.

Martin, Emily. 1989. *The Woman in the Body*. Milton Keynes: Open University Press.

Martin, Emily. 1991. "The Egg and the Sperm: How Science Has Constructed a Romance Based on Stereotypical Male-Female Roles." *Signs* 16(3): 485–501.

Martin, Emily. 1995. *Flexible Bodies: The Role of Immunity in American Culture from the Days of Polio to the Age of AIDS*. Boston: Beacon Press.

Matchar, Emily. 2013. *Homeward Bound: Why Women Are Embracing the New Domesticity*. New York: Simon and Schuster.

Mauldin, Laura. 2012. "Parents of Deaf Children with Cochlear Implants: A Study of Technology and Community." *Sociology of Health and Illness* 34(4): 529–43.

McBean, A. M., Li, S., Gilbertson, D. T., and Collins, A. J. 2004. "Differences in Diabetes Prevalence, Incidence, and Mortality among the Elderly of Four Racial/ethnic Groups: Whites, Blacks, Hispanics, and Asians." *Diabetes Care* 27(10): 2317–24.

McDonough, Katie. 2013. "Israel Admits Ethiopian Jewish Immigrants Were Given Birth Control Shots." *Salon*, January 28. To access: http://www.salon.com/2013/01/28/israel_admits_ethiopian_jewish_immigrants_were_given_birth_control_shots/.

McGlotten, Shaka and Lisa Jean Moore. 2013. "Dry and Limp: Aging Queers, Zombies, and Sexual Reanimation." *Journal of Medical Humanities*. 34(2): 261–268.

McIntosh, Peggy. 1988. "White Privilege and Male Privilege: A Personal Account of Coming to See Correspondences through Work in Women's Studies." *Working Paper #189*, Wellesley College Center for Research on Women, Wellesley, MA 02181.

Mead, George Herbert. 1934. *Mind, Self, and Society*. Chicago: University of Chicago Press.

Mello, Susan. 2012. "Selling a Super Cosmeceutical: Contextualizing Risk in Direct-to-Consumer Advertising of BOTOX(r) Cosmetic." *Health, Risk and Society* 14(4): 385–98.

Micheva, K. D., B. Busse, N. C. Weiler, N. O'Rourke, and S. J. Smith. 2010. "Single-Synapse Analysis of a Diverse Synapse Population: Proteomic Imaging Methods and Markers." *Neuron* 68(4): 639–53.

Mientka, Matthew. 2013. "Poor, Minority Women Most Likely to Die Young from Breast Cancer." *Medical Daily*, April 29.

Miller, Toby. 2005. "A Metrosexual Eye on Queer Guy. GLQ." *A Journal of Lesbian and Gay Studies* 11(1): 112–17.

Mills, C. Wright. 1959. *The Sociological Imagination*. London: Oxford University Press.

Modell, Bernadette and Matthew Darlison. 2008. "Global Epidemiology of Haemoglobin Disorders and Derived Service Indicators." *Bulletin of the World Health Organization* 86(6). http://www.scielosp.org/scielo.php?pid=S0042-96862008000600017&script=sci_arttext.

Mojtehedzadeh, Sara. 2013. "The Throwaways." *Guernica*, March 1. To access: http://www.guernicamag.com/features/the-throwaways/.

Moniruzzaman, Monir. 2012. "'Living Cadavers' in Bangladesh: Bioviolence in the Human Organ Bazaar." *Medical Anthropology Quarterly* 26(1): 69–91.

Moore, Davis S. 2003. *The Dependent Gene: The Fallacy of "Nature vs. Nurture."* New York: Holt Paperbacks.

Moore, Lisa Jean. 2007a. *Sperm Counts: Overcome by Man's Most Precious Fluid.* New York: New York University Press.

Moore, Lisa Jean. 2007. "Incongruent Bodies: Teaching While Leaking." *Feminist Teacher* 17(2): 1–12.

Moore, Lisa Jean and Adele E. Clarke. 1995. "Clitoral Conventions: Graphic Representations of Female Genital Anatomy, c 1900–1991." *Feminist Studies* 21(2): 255–301.

Moore, Lisa Jean and Juliana Weissebein. 2010. "Cocktail Parties: Fetishizing Semen in Pornography Beyond Bukkake." In Karen Boyle, ed. *Everyday Pornographies*. London: Routledge, pp. 77–89.

Moore, Lisa Jean and Marianna Grady. 2014. "Putting Daddy in the Cart: Ordering Sperm Online." In Meredith Nash, ed. *Reframing Reproduction: Sociological Perspective on the Intersection between Gender, Sexuality and Reproduction in Late Modernity*. New York: Palgrave-McMillan.

Moreton-Robinson, Aileen. 2002. *Talkin' Up to the White Woman: Indigenous Women and Feminism.* St Lucia, QLD: University of Queensland Press.

Morgan, Lynn M. 2009. *Icons of Life: A Cultural History of Human Embryos.* Berkeley: University of California Press.

Morgan, Susan E., Tyler R. Harrison, Walid A. Afifi, Shawn D. Long, and Michael T. Stephenson. 2008. "In Their Own Words: The Reasons Why People Will (Not) Sign an Organ Donor Card." *Health Communication* 23: 23–33.

Morning, Ann. 2011. *The Nature of Race: How Scientists Think and Teach About Human Difference*. Berkeley: University of California Press.

Morrison, Daniel R. and Monica J. Casper. 2011. "Does Traumatic Brain Injury Have a Gender?" *The Feminist Wire*, February 22. http://thefeministwire.com/2011/02/does-traumatic-brain-injury-have-a-gender/.

Morrison, Daniel R. and Monica J. Casper. 2012. "Intersections of Disability Studies and Critical Trauma Studies: A Provocation." *Disability Studies Quarterly* 32(2).

Morsy, Soheir A. 1986. "Deadly Reproduction Among Egyptian Women: Maternal Mortality and the Medicalization of Population Control." pp. 162–76 in Faye D. Ginsburg and Rayna Rapp. eds. *Conceiving the New World Order: The Global Politics of Reproduction.* Berkeley: University of California Press.

Moshenska, Gabriel. 2010. "Gas Masks: Material Culture, Memory, and the Senses." *Journal of the Royal Anthropological Institute* 16(3): pp. 609–28.

Motoc, Iulia Voina. 2008 "The International Law of Genetic Discrimination: The Power of 'Never Again.'" *Center for Human Rights and Global Justice Working Paper* 18: 222–45.

Moynihan, Ray. 2003. "The Making of a Disease: Female Sexual Dysfunction." *British Medical Journal*. 326: 45–47.

Mykytyn, Courtney Everts. 2010. "A History of the Future: The Emergence of Contemporary Anti-Ageing Medicine." *Sociology of Health and Illness* 32(2): 181–96.

Nelkin, D. and Lindee, M. S. 1995. "The DNA Mystique: The Gene as a Cultural Icon." New York: WH Freeman and Co.

Nelson, Alondra. 2011. *Body and Soul: The Black Panther Party and the Fight Against Medical Discrimination*. Minneapolis: University of Minnesota Press.

Nielsen, Jared A., Brandon A. Zielinski, Michael A. Ferguson, Janet E. Lainhart, and Jeffrey S. Anderson. 2013. "An Evaluation of the Left-Brain vs. Right-Brain Hypothesis with Resting State Functional Connectivity Magnetic Resonance Imaging." *PLoS One*, August 14.

Niewenhuys, Rudolf, Jan Voogd, and Christiaan van Huijzen. 2007. *The Human Central Nervous System: A Synopsis and Atlas*. 4th edition. Dresden, Germany: Steinkopff.

Nijhuis, Michelle. 2008. "Friend or Foe? Crows Never Forget a Face, It Seems." *New York Times*, August 25.

Niland, Patricia and Antonia C. Lyons. 2011. "Uncertainty in Medicine: Meanings of Menopause and Hormone Replacement Therapy in Medical Textbooks." *Social Science and Medicine* 73(8): 1238–45.

Nisbet, Matthew C. and Declan Fahy. 2013. "Bioethics in Popular Science: Evaluating the Media Impact of *The Immortal Life of Henrietta Lacks* on the Biobank Debate." *BMC Medical Ethics* 14: 10.

Noren, Torbjorn. 2010. "Clostridium Difficile and the Disease It Causes." *Methods in Molecular Biology* 646: 9–35.

Nussbaum, Martha C. and Cass R. Sunstein, eds. 1999. *Clones and Clones: Fact and Fantasies About Human Cloning*. New York: W.W. Norton.

O'Farrell, N. and M. Egger. 2000 "Circumcision in Men and the Prevention of HIV Infection: A Meta-analysis Revisited." *International Journal of Sexually Transmitted Diseases and AIDS* 11: 137–42.

O'Neill, Kate. 2003. "A Vital Fluid: Risk, Controversy, and the Politics of Blood Donation in the Era of 'Mad Cow Disease.'" *Public Understanding of Science* 12(4): 359–80.

Oettgen, Peter. 2006. "Cardiac Stem Cell Therapy: Need for Optimization of Efficacy and Safety Monitoring." *Circulation* 114: 353–58.

Okabi, Yasser. 2013. "Syrian Rebels Making Homemade Gas Masks." *The Jerusalem Post*. accessed http://www.jpost.com/Middle-East/Syrian-rebels-making-home made-gas-masks-311288.

Olumide, Y. M., Akinkugbe, A. O., Altraide, D., Mohammed, T., Ahamefule, N., Ayanlowo, S., Onyekonwu, C. and Essen, N. 2008. "Complications of Chronic

Use of Skin Lightening Cosmetics." *International Journal of Dermatology* 47: 344–53.
Omi, Michael and Howard Winant. 1994. *Racial Formations in the United States: From the 1960s to the 1990s*. 2nd edition. New York: Routledge.
Oshinsky, David. 2006. *Polio: An American Story*. New York: Oxford University Press.
Oudshoorn, Nelly. 1994. *Beyond the Natural Body: An Archaeology of Sex Hormones*. London: Routledge.
Oudshoorn, Nelly. 2003. *The Male Pill: A Biography of a Technology in the Making*. Durham, NC: Duke University Press.
Padamsee, Tasleem Juana. 2011. "The Pharmaceutical Corporation and the 'Good Work' of Managing Women's Bodies." *Social Science and Medicine* 72(8): 1342–50.
Padania, Sameer. 2012. "How to Defend Yourself Against Facial Recognition Technology." *Mediashift*, June 18. http://www.pbs.org/mediashift/2012/06/how-to-defend-yourself-against-facial-recognition-technology170/.
Paltrow, Lynn M. and Jeanne Flavin. 2013. "Arrests of and Forced Interventions on Pregnant Women in the United States, 1973–2005; Implications for Women's Legal Status and Public Health." *Journal of Health Politics, Policy and Law* 38(2): 299–343.
Parens, Erik and Adrienne Asch. 1999. "The Disability Rights Critique of Prenatal Genetic Testing: Reflections and Recommendations." *The Hastings Center Report* 29(5): 1–22.
Parr, Lisa A. 2011. "The Evolution of Face Processing in Primates." *Philosophical Transactions of the Royal Society of the Biological Sciences* 366: 1571, 1764–77.
Parsons, Talcott. 1951a. *The Social System*. London: The Free Press.
Parsons, Talcott. 1951b. "Illness and the Role of the Physician: A Sociological Perspective." *American Journal of Orthopsychiatry* 21(3): 452–60.
Parsons, Talcott. 1964. *Social Structure and Personality*. New York: The Free Press.
Patton, Cindy. 1990. *Inventing AIDS*. New York: Routledge.
Patz, Jonathan, Diarmid Campell-Lendrum, Tracey Holloway and Jonathan Foley. 2005. "Impact of Regional Climate Change on Human Health." *Nature* 438: 310–17.
Paus, R. and Cotsarelis, G. 1999. "The Biology of Hair Follicles." *New England Journal of Medicine* 341(7): 491–97.
Pavoni, Vittorio, Lara Gianesello, Laura Paparella, Laura Tadini Buoninsegni, and Elisabetta Barboni. 2010. "Outcome Predictors and Quality of Life of Severe Burn Patients Admitted to Intensive Care Unit." *Scandinavian Journal of Trauma, Resuscitation, and Emergency Medicine* 18: 24.
Payer, Lynn. 1992. *Disease-mongers: How Doctors, Drug Companies, and Insurers are Making you Feel Sick*. New York: Wiley and Sons.
PBS. 2002. "Red Gold: The Epic Story of Blood." http://www.pbs.org/wnet/redgold/about/index.html.
Peltola, Heikki. 2001. "Burden of Meningitis and Other Severe Bacterial Infections of Children in Africa: Implications for Prevention." *Clinical Infectious Diseases* 32(1): 64–75.

Peters, Stella Roos. 2012. "Invisible Victims: The Effects of Structural Violence on Infant and Child Mortality in Papua Barat, Indonesia in the Context of Human Rights." MA Thesis in Conflict Studies and Human Rights, Utrecht University, August 20. To access: http://www.invisiblevictims.nl/uploads/7/6/9/7/7697426/invisible_victims.pdf.

Petryna, Adriana, Andrew Lakoff, and Arthur Kleinman, eds. 2006. *Global Pharmaceuticals: Ethics, Markets, Practices*. Durham, NC: Duke University Press.

Phillips, W. M. 1993. "The Artificial Heart: History and Current Status." *Journal of Biomechanical Engineering* 115(4B): 555–57.

Pilgrimed.com, "Cost Comparisons." To access: http://pilgrimed.com/medical-tourism-information/cost-comparisons.html.

Pitts-Taylor, Victoria. 2007. *Surgery Junkies: Wellness and Pathology in Cosmetic Culture*. New Brunswick, NJ: Rutgers University Press.

Plotnik, Joshua M., Frans B. M. de Waal, and Reiss. 2006. "Self-Recognition in an Asian Elephant." *Proceedings of the National Academy of Sciences* 103(45): 17053–57.

Pollen, Michael. 2013. "Some of My Best Friends Are Germs." *New York Times*, May 15.

Polsky, Allyson D. 2002. "Blood, Race, and National Identity: Scientific and Popular Discourses." *Journal of Medical Humanities* 23(3/4): 171–86.

Porter, Peggy. 2008. "'Westernizing' Women's Risks: Breast Cancer in Lower-Income Countries." *New England Journal of Medicine* 358: 213–16. To access: http://www.nejm.org/doi/full/10.1056/nejmp070830.

Power, D'Arcy. 2005. *William Harvey: Masters of Medicine*. White fish, Montana: Kessinger Publishing.

Pray, Leslie A. 2008. "Embryo Screening and the Ethics of Human Genetic Engineering." *Nature Education* 1(1): 207.

Preves, Sharon E. 2003. *Intersex and Identity: The Contested Self.* New Brunswick, NJ: Rutgers University Press.

Proctor, Robert. 1988. *Racial Hygiene: Medicine Under the Nazis*. Cambridge, MA: Harvard University Press.

Public Citizen. 2008. "Stricter Warnings Needed for Botox, Myobloc Injections." Press release, January 24. To access: http://www.citizen.org/pressroom/pressroomredirect.cfm?ID=2593.

Pullen, P. R., W. R. Thompson, D. Benardot, L. J. Brandon, P. K. Mehta, L. U. A. Y. Rifai and B. V. Khan. 2010. "Benefits of Yoga for African American Heart Failure Patients." *Med Sci Sports Exerc* 42(4): 651–57.

Quart, Alissa. 2012. "Neuroscience: Under Attack." *New York Times*, November 23.

Rabin, Roni Caryn. 2009. "Officials Weigh Circumcision to Fight HIV Risk." *The New York Times*, August 24. Retrieved from http://www.nytimes.com/2009/08/24/health/policy/24circumcision.html.

Rabin, Roni Caryn. 2009. "The Latest Fight Over the Foreskin." *The New York Times*, 29 August. To access: http://www.nytimes.com/2009/08/30/weekinreview/30rabin.html.

Ramsey, Sara., C. Sweeney, M. Fraser, and G. Oades. 2009. "Pubic Hair and Sexuality: A Review." *Journal of Sexual Medicine* 6: 2102–10.
Ramsay, S. E., P. H. Whincup, L. T. Lennon, R. W. Morris and S. G. Wannamethee. 2011. "Longitudinal Associations of Socioeconomic Position in Childhood and Adulthood with Decline in Lung Function over 20 Years: Results from a Population-based Cohort of British Men." *Thorax* 66(12): 1058–64.
Rapp, Rayna. 2000. *Testing Women, Testing the Fetus: The Social Impact of Amniocentesis in America*. New York: Routledge.
Reaves, Jessica. 2001. "The Great Debate Over Stem Cell Research." *Time*, July 11.
Reis, Elizabeth. 2012. *Bodies in Doubt: An American History of Intersex*. Baltimore: Johns Hopkins University Press.
Reiss, Diana and Lori Marino. 2001. "Mirror Self-Recognition in the Bottlenose Dolphin." *Proceedings of the National Academy of Sciences* 98(10): 5937–42.
Rhodes, T. et al. 1998. "Prevalence of Male Pattern Hair Loss in 18–49 Year Old Men." *Dermatologic Surgery* 24(12): 1330–32.
Rich, Adrienne. 1976. *Of Woman Born: Motherhood as Experience and Institution*. New York: Norton.
Richard, Michael Graham. 2013. "Fact: Air Pollution Kills More People than AIDS and Malaria Combined." *Mother Nature Network*. accessed http://www.mnn.com/health/fitness-well-being/stories/fact-air-pollution-kills-more-people-than-aids-and-malaria.
Richardson, Sarah S. 2013. *Sex Itself: The Search for Male and Female in the Human Genome*. Chicago: University of Chicago Press.
Roberts, Dorothy. 1998. *Killing the Black Body: Race, Reproduction, and the Meaning of Liberty*. New York: Vintage.
Roberts, Dorothy. 2012. *Fatal Invention: How Science, Politics, and Big Business Re-Create Race in the Twenty-First Century*. New York: The New Press.
Rogers, Benjamin A., Zohreh Aminzadeh, Yoshiro Hayashi, and David L. Paterson. 2011. "Country-to-Country Transfer of Patients and the Risk of Multi-Resistant Bacterial Infection." *Clinical Infectious Diseases* 53(1): 49–56.
Rogowska-Szadkowska, D. 2011. "Consequences of the Commercialization of Plasma and Blood in China." *Przegl Epidemiology* 65(3): 515–19.
Rondilla, Joanne and Paul Spickard. 2007. *Is Lighter Better? Skin-tone Discrimination among Asian Americans*. Lanham, MD: Rowman & Littlefield.
Rose, Nikolas. 2007. *The Politics of Life Itself: Biomedicine, Power, and Subjectivity in the Twenty-First Century*. Princeton: Princeton University Press.
Rose, Nikolas. 2012. "The Human Sciences in a Biological Age." *Institute for Culture and Society Occasional Paper* 3(1): 1–24.
Rosenberg, Mark W. and Kathleen Wilson. 2000. "Gender, Poverty and Location: How Much Difference Do They Make In the Geography of Health Inequalities?" *Social Science and Medicine* 51(2): 275–87.
Roskies, Adina L. 2002. "Neuroethics for the New Millennium." *Neuron* 35: 21–23.
Rosman, Sophia. 2004. "Cancer and Stigma: Experiences of Patients with Chemotherapy Induced Alopecia." *Patient Education and Counseling* 52(3): 333–39.

Roth, Rachel. 2003. *Making Women Pay: The Hidden Costs of Fetal Rights*. Ithaca, NY: Cornell University Press.
Rothblum, Esther and Sondra Solovay. 2009. *The Fat Studies Reader*. New York: NYU Press.
Rothman, Barbara Katz. 1989. *Recreating Motherhood*. New York: W.W. Norton & Co.
Rothman, Barbara Katz. 2005. *Weaving a Family: Untangling Race and Adoption*. Boston: Beacon Press.
Rubin, Gayle. 1975. "The Traffic in Women: Notes on the 'Political Economy' of Sex." In Rayna Reiter, ed. *Toward an Anthropology of Women*. New York: Monthly Review Press.
Runtstedtler, Theresa. 2012. Jack Johnson, *Rebel Sojourner: Boxing in the Shadow of the Global Color Line*. Berkeley: University of California Press.
Ruzek, Sheryl Burt. 1978. *The Women's Health Movement: Feminist Alternatives to Medical Control*. New York: Praeger.
Ryan, Donna F. and John S. Schuchman, eds. 2002. *Deaf People in Hitler's Europe*. Washington, DC: Gallaudet University Press.
Saba, M. M., H. O. Ventura, M. Saleh, and M. R. Mehra. 2006. "Ancient Egyptian Medicine and the Concept of Heart Failure." *Journal of Cardiac Failure* 12(6): 416–21.
Saguy, Abigail C. 2013. *What's Wrong with Fat?* New York: Oxford University Press.
Saleem, Shabana and Martin Bobak. 2005. "Women's Autonomy, Education, and Contraception Use in Pakistan: A National Study." *Reproductive Health* 2(8).
Saletan, William. 2009. "Sex Reversal: Child Quotas, Abortion, and China's Missing Girls." *Slate*, April 15.
Samimi, Jennifer Ceema. 2010. "Funding America's Nonprofits: The Nonprofit Industrial Complex's Hold on Social Justice." *Columbia Social Work Review*, Volume I. http://cswr.columbia.edu/wp-content/uploads/2012/10/2010_vol1_pg17_samimi.pdf.
Sanders, Clinton. 2008. *Customizing the Body: The Art and Culture of Tattooing*. Philadelphia: Temple University Press.
Satel, Sally and Scott O. Lilienfeld. 2013. *Brainwashed: The Seductive Appeal of Mindless Neuroscience*. New York: Basic Books.
Saugstad, Ola Didrik and Torleiv Ole Rognum. 2003. "Sudden Infant Death Syndrome is Preceded by Hypoxia." *Pediatric Research* 53: 881–82.
Schaefer, Andrew M., Robert W. Taylor, Douglass M. Turnbull, and Patrick F. Chinnery. 2004. "The Epidemiology of Mitochondrial Disorders: Past, Present, and Future." *BBA Bioenergetics* 2–3(1659): 115–20.
Schafer, Markus H. and Kenneth F. Ferraro. 2011. "The Stigma of Obesity: Does Perceived Weight Discrimination Affect Identity and Physical Health?" *Social Psychology Quarterly* 74(1): 76–97.
Scheper-Hughes, Nancy. 2011. "Mr. Tati's Holiday and João's Safari: Seeing the World Through Transplant Tourism." *Body and Society* 17(2): 55–92.
Schettler, Ted, Gina Solomon, Maria Valenti, and Annette Huddle. 1999. *Generations At Risk: Reproductive Health and the Environment*. Cambridge, MA: MIT Press.

Schiebinger, Londa. 1993. "Why Mammals Are Called Mammals: Gender Politics in Eighteenth-Century Natural History." *The American Historical Review* 98(2): 382–411.
Schultz, K. T. and F. Grieder. 1987. "Structure and Function of the Immune System." *Toxicologic Pathology* 15(3): 262–64.
Sedgwick, Eve. 1991. "How to Bring Your Kids up Gay." *Social Text* 29: 18–27.
Seigworth, Gregory J. and Melissa Gregg. 2010. "An Inventory of Shimmers." pp. 1–25 in *The Affect Theory Reader*, edited by Melissa Gregg and Gregory J. Seigworth. Durham, NC: Duke University Press.
Selander, Jodi, Allison Cantor, Sharon M. Young, and Daniel C. Benyshek. 2013. "Human Maternal Placentophagy: A Survey of Self-Reported Motivations and Experiences Associated with Placenta Consumption." *Ecology of Food and Nutrition* 52: 93–115.
Sesin, Carmen. 2013. "Despite Strong Warnings, Families Turn to Latin America for Stem Cell 'Tourism.'" *NBC Latino*, September 26. http://nbclatino.com/2013/09/26/despite-strong-warnings-families-turn-to-latin-america-for-stem-cell-tourism/.
Severinsen, Stig Avall 2010. *Breatheology*. Naples, Italy: Idelson Gnocchi Publishers.
Shaw, J. E., Sicree, R. A. and P. Z. Zimmet. 2010. "Global Estimates of the Prevalence of Diabetes for 2010 and 2030." *Diabetes Research and Clinical Practice* 87(1) 4–14.
Shea, Beth. 2009. "Placenta Teddy Bear." *Inhabitots.com*, October 1. Access: http://www.inhabitots.com/doing-it-for-the-kids-design-exhibition-placenta-teddy-bear/.
Shepherd, R., J. Barnett, H. Cooper, A. Coyle, J. Moran-Ellis, V. Senior, and C. Walton. 2007. "Towards an Understanding of British Public Attitudes Concerning Human Cloning." *Social Science and Medicine* 65(2): 377–92.
Sherrow, Victoria. 2006. *Encyclopedia of Hair: A Cultural History*. Santa Barbara, CA: Greenwood.
Shildrick, Margrit. 1997. *Leaky Bodies and Boundaries: Feminism, Postmodernism, and (Bio)Ethics*. London: Routledge.
Shildrick, Margrit and Janet Price. 1999. "Openings on the Body: A Critical Introduction." In Price, Janet and Margrit Shildrick (eds.) *Feminist Theory and The Body: A Reader*. New York: Routledge. 1–14.
Shilling, Chris. 1993. *The Body and Social Theory*. 2nd edition London: Sage.
Shimazono, Yosuke. 2007. "The State of the International Organ Trade: A Provisional Picture Based on Integration of Available Information." *Bulletin of the World Health Organization* 85(12), 901–980.
Shiva, Vandana. 2000. *Stolen Harvest: The Hijacking of the Global Food Supply*. Boston: South End Press.
Silk, Michael L., David L. Andrews, and C. L. Cole, eds. 2005. *Sport and Corporate Nationalisms*. London: Bloomsbury Academic.
Silliman, Jael, Marlene Gerber Fried, Loretta Ross, and Elena R. Gutierrez, eds. 2004. *Undivided Rights: Women of Color Organizing for Reproductive Justice*. Boston: South End Press.

Silverstein, Charles. 2009. "The Implications of Removing Homosexuality from the *DSM* as a Mental Disorder." *Archives of Sexual Behavior* 38: 161–63.

Simón-Sánchez, Javier and Andrew Singleton. 2008. "Genome-Wide Association Studies in Neurological Disorders." *Lancet Neurology* 7(11): 1067–72.

Simpson, J. L. 2007. "Causes of Fetal Wastage." *Clinical Obstetrics and Gynecology* 50(1): 10–30.

Sinues, P., M. Kohler and R. Zenobi. 2013. "Human Breath Analysis may support the existence of individual metabolic phenotypes." PLoS. http://www.plosone.org/article/info%3Adoi%2F10.1371%2Fjournal.pone.0059909.

Sion, Brigitte. 2011. "Conflicting Sites of Memory in Post-Genocide Cambodia." *Humanity: An International Journal of Human Rights* 2(1): 1–21.

Sisjord, Mari Kristin and Elsa Kirstiansen. 2009. "Elite Women Wrestlers' Muscles: Physical Strength and a Social Burden." *International Review for the Sociology of Sport* 44 (2–3): 231–46.

Skloot, Rebecca. 2011. *The Immortal Life of Henrietta Lacks*. New York: Broadway Books.

Slate. 2001. "How do Terrorist 'Cells' Work?" September 17. http://www.slate.com/articles/news_and_politics/explainer/2001/09/how_do_terrorist_cells_work.html.

Smith, Barbara Ellen. 1987. *Digging Our Own Graves: Coal Miners and the Struggle over Black Lung Disease*. Philadelphia: Temple University Press.

Smith, James. 2007. "Diabetes and the Rise of the SES Health Gradient." NBER Working Paper No. 12905.

Solovay, Sondra and Esther Rothblum. 2009. "Introduction." In Esther Rothblum and Sondra Solovay, eds. *The Fat Studies Reader*. New York: NYU Press.

Spooner, J. 1967. "History of Surgical Face Masks: The Myths, the Masks, and the Men and Women behind Them." *AORN* 5(1): 76–80.

Stein, Ed. 2001. *The Mismeasure of Desire: The Science, Theory and Ethics of Sexual Orientation*. Oxford: Oxford University Press.

Stein, Elissa and Susan Kim. 2009. *Flow: The Cultural Story of Menstruation*. New York: St. Martin's Griffin.

Stein, Rob. 2013. "Getting Your Microbes Analyzed Raises Big Privacy Issues." NPR, November 4. http://www.npr.org/blogs/health/2013/11/04/240278593/getting-your-microbes-analyzed-raises-big-privacy-issues.

Steiner, Leslie Morgan. 2007. *Mommy Wars: Stay-At-Home and Career Moms Face Off on Their Choices, Their Lives, Their Families*. New York: Random House.

Steingraber, Sandra. 2001. *Having Faith: An Ecologist's Journey to Motherhood*. New York: Perseus Publishing.

Stern, Alexandra Minna. 2005. *Eugenic Nation: Faults and Frontiers of Better Breeding in Modern America*. Berkeley: University of California Press.

Strudwick, Helen. 2006. *The Encyclopedia of Ancient Egypt*. Singapore: De Agostini UK.

Stryker, Susan. 2008. *Transgender History*. Berkeley, CA: Seal Press.

Styrker, Susan and Aren Aizura. 2013. *The Transgender Studies Reader* 2. New York: Routledge.

Sulik, Gayle. 2012. *Pink Ribbon Blues: How Breast Cancer Culture Undermines Women's Health*. New York: Oxford University Press.
Surbone, A. 2011. "Social and Ethical Implications of BRCA Testing." *Annals of Oncology* 22(Suppl): 60–66.
Sy, Stephanie. 2009. "More Than a Game: Testing Your Blood Alcohol." *ABC News*. http://abcnews.go.com/WN/people-personal-breathalyzers-avoid-driving-drunk/story?id=9431050#.UbX1UGTEpi4.
Sykes, Bryan. 2003. *Adam's Curse: A Future without Men*. New York: Bantam.
Synnott, Anthony. 1987. "Shame and Glory: A Sociology of Hair." *British Journal of Sociology* 38: 381–413.
Szabo, R. and R. V. Short. 2000. "How Does Male Circumcision Protect Against HIV Infection?" *British Medical Journal* 320: 1592–94.
Szasz, Andrew. 2007. *Shopping Our Way to Safety: How We Changed from Protecting the Environment to Protecting Ourselves*. Minneapolis: University of Minnesota Press.
Tack, G. J., Verbeek, W. H., Schreurs, M. W., and Mulder, C. J. 2010. "The Spectrum of Celiac Disease: Epidemiology, Clinical Aspects and Treatment." *Nature Reviews Gastroenterology and Hepatology* 7(4): 204–13.
Tan, Arian James and Doug Brignole. 2012. *Million Dollar Muscle: A Historical and Sociology Perspective on the Fitness Industry*. New York: Cognella Academic Publishing.
Taylor, Janelle S. 2008. *The Public Life of the Fetal Sonogram: Technology, Consumption, and the Politics of Reproduction*. New Brunswick, NJ: Rutgers University Press.
Teixeira, M. Elizabeth and Geraldine M. Budd. 2010. "Obesity Stigma: A Newly Recognized Barrier to Comprehensive and Effective Type 2 Diabetes Management." *Journal of the American Academy of Nurse Practitioners* 22(10): 527–33.
Terry, Jennifer. 2009. "Significant Injury: War, Medicine, and Empire in Claudia's Case." *WSQ: Women's Studies Quarterly* 37(1&2): 200–25.
Thacker, Eugene. 2005. *The Global Genome: Biotechnology, Politics, and Culture*. Cambridge, MA: MIT Press.
Tiefer, Leonore. 2006. "Female Sexual Dysfunction: A Case Study of Disease Mongering and Activist Resistance." *PLoS Medicine* 3(4): 436–40.
Tillisch, Kirsten, et al. 2013. "Consumption of Fermented Milk Product with Probiotic Modulates Brain Activity." *Gastroenterology* 144: 1394–401.
Timmermans, Stefan and Steven Haas. 2008. "Toward a Sociology of Disease." *Sociology of Health and Illness* 30(5): 659–76.
Timmermans, Stefan and Mara Buchbinder. 2012. *Saving Babies? The Consequences of Newborn Genetic Screening*. Chicago: University of Chicago Press.
Todorova, I., A. Alexandrova-Karamanova, and J. Bradley. 2009. "Inequalities in Cervical Cancer Screening in Eastern Europe: Perspectives from Bulgaria and Romania." *International Journal of Public Health* 54(4): 222–32.
Toilet Training. 2003. Directed by Tara Mateik and the Sylvia Rivera Law Project. DVD. New York: Sylvia Rivera Law Project.

Tomasi, Dardo and Nora D. Volkow. 2012. "Laterality Patterns of Brain Functional Connectivity: Gender Effects." *Cerebral Cortex* 22(6): 1455–62.

Tone, Andrea. 2002. *Devices and Desires: A History of Contraceptives in America*. 1st edition. New York: Hill and Wang.

Treichler, Paula. 1999. *How to Have a Theory in an Epidemic: Cultural Chronicles of AIDS*. Durham, NC: Duke University Press.

Truog, Robert D., Aaron S. Kesselheim, and Steven Joffe. 2012. "Paying Patients for their Tissue: The Legacy of Henrietta Lacks." *Science* 337(July 6): 37–38.

Tsoffar, Ruth. 2004. "The Body as Storyteller: Karaite Women's Experience of Blood and Milk." *Journal of American Folklore* 117 (463): 3–21.

Turner, Bryan. 1984. *The Body and Society*. Oxford: Basil Blackwell.

Turner, Bryan. 1999. "The Possibility of Primitiveness: Towards a Sociology of Body Marks in Cool Societies." *Body & Society* 5(2–3): 39–50.

Twine, France Winddance. 1998. *Racism in a Racial Democracy: The Maintenance of White Supremacy in Brazil*. New Brunswick, NJ: Rutgers University Press.

Twine, France Winddance. 2011. *Outsourcing the Womb: Race, Class and Gestational Surrogacy in a Global Market*. New York: Routledge.

Udry, J. Richard. 1995. "Sociology and Biology: What Biology Do Sociologists Need to Know?" *Social Forces* 73(4): 1267–78.

van den Wijngaard, Marianne. 1997. *Reinventing the Sexes: The Biomedical Construction of Femininity and Masculinity*. Bloomington: Indiana University Press.

Vannini, Phillip and Aaron M. McCright. 2004. "To Die For: The Semiotic Seductive Power of the Tanned Body." *Symbolic Interaction* 27(3): 309–32.

Vesely, Rebecca. 2003. "Study: Women Bear Brunt of Environmental Toxins." *Women's E-News*, October 21.

Victor, Christina, Sasha Scambler and John Bond. 2008. *The Social World of Older People: Understanding Loneliness and Social Isolation in Later Life*. Maidenhead: Open Society Press.

Villarreal, Andrés. 2010. "Stratification by Skin Color in Contemporary Mexico." *American Sociological Review* 75: 652–78.

Virtual Medical Center. 2012. "Liquid Gold: Breast Milk Banking." http://www.virtualmedicalcentre.com/news/liquid-gold-breast-milk-banking/17224 accessed on August 27, 2013.

Wald, Priscilla. 2008. *Contagious: Cultures, Carriers, and the Outbreak Narrative*. Durham, NC: Duke University Press.

Waldby, Catherine and Robert Mitchell. 2006. *Tissue Economies: Blood, Organs, and Cell Lines in Late Capitalism*. Durham, NC: Duke University Press.

Walters, LeRoy. 2004. "Human Embryonic Stem Cell Research: An Intercultural Perspective." *Kennedy Institute of Ethics Journal* 14(1): 3–38.

Wang, Ming-Hsien and Laurence Baskin. 2008. "Endocrine Disruptors, Genital Development and Hypospadias." *Journal of Andrology* 29(5):499–505.

Warren, Michael. 2007. "Interpreting Gunshots in Wounds in the Balkans: Evidence for Genocide." In Brickley, Megan and Roxana Ferllini (eds.) *Forensic Anthropology: Case Studies from Europe*. New York: Charles C. Thomas. 151–64.

Watkins, Elizabeth Siegel. 2001. *On the Pill: A Social History of Oral Contraceptives, 1950–1970*. Baltimore: The Johns Hopkins University Press.
Weisberg, Deena Skolnick, Frank C. Keil, Joshua Goodstein, Elizabeth Rawson, and Jeremy R. Gray. 2008. "The Seductive Allure of Neuroscience Explanations." *Journal of Cognitive Neuroscience* 20(3): 470–77.
Weiss, H. A., M. A. Quigley, and R. J. Hayes. 2000. "Male Circumcision and Risk of HIV Infection in Sub-Saharan Africa: A Systematic Review and Meta-Analysis." *AIDS* 14 (Oct 20): 2361–67.
Weisz, George, ed. 1991. *Social Science Perspectives on Medical Ethics*. Philadelphia: University of Pennsylvania Press.
Weitz, Rose. 2004. *Rapunzel's Daughters: What Women's Hair Tells Us about Women's Lives*. New York: Farrar, Straus and Giroux.
Wenger, Nanette K. 2002. "Clinical Characteristics of Coronary Heart Disease in Women: Emphasis on Gender Differences." *Cardiovascular Research* 53: 558–67.
Wenger, Nanette K., Leon Speroff, and Barbara Packard. 1993. "Cardiovascular Health and Disease in Women." *New England Journal of Medicine* 329: 247–56.
West, John B. 2008. "Ibn al-Nafis, the Pulmonary Circulation, and the Islamic Golden Age." *Journal of Applied Physiology* 105(6): 1877–80.
Wick, Jeannette Yeznach. 2012. "Lou Gehrig's Disease." *Pharmacy Times*, February. http://www.pharmacytimes.com/media/pdf/3bbcbcc203b370efc906310041b12a43.pdf.
Wilgus, Traci A. 2007. "Regenerative Healing in Fetal Skin: A Review of the Literature." *Ostomy Wound Management* 53(6): 16–31.
Williams, Johnny E. 2011. "They Say It's in the Genes: Decoding Racial Ideology in Genomics." *Journal of Contemporary Ethnography* 40: 550–81.
Williams, Simon and Gillian Bendelow. 1998. *The Lived Body: Sociological Themes, Embodied Issues*. New York: Routledge.
Wilson, Edward O. 1975. *Sociobiology: The New Synthesis*. Cambridge, MA: Harvard University Press.
Wilson, Edward O. 1979. *On Human Nature*. Cambridge, MA: Harvard University Press.
Wilson, Peter W. F., Ralph B. D'Agostino, Daniel Levy, Albert M. Belanger, Halit Silbershatz, and William B. Kannel. 1998. "Prediction of Coronary Heart Disease Using Risk Factor Categories." *Circulation* 97: 1837–47.
Wolf, Joan. 2010. *Is Breast Best? Taking on the Breastfeeding Experts and the New High Stakes of Motherhood*. New York: NYU Press.
Wolf, Naomi. 2002 [1991]. *The Beauty Myth: How Images of Beauty Are Used Against Women*. New York: Perennial.
World Health Organization. 2008. *Male Circumcision: Global Trends and Determinants of Prevalence, Safety and Acceptability*. World Health Organization.
World Health Organization and Alzheimer's Disease International. 2012. *Dementia: A Public Health Priority*. Geneva: WHO.
Wright, Anne L. and Richard J. Schanler. 2001. "The Resurgence of Breastfeeding at the End of the Second Millennium." *The Journal of Nutrition* 131(2): 421S–425S.

Wyndaele, M. and J.-J. Wyndaele. 2006. "Incidence, Prevalence, and Epidemiology of Spinal Cord Injury: What Learns A Worldwide Literature Survey?" *Spinal Cord* 44: 508–23.

Xu, F., L. E. Markowitz, M. R. Sternberg and S. O. Aral 2007. "Prevalence of Circumcision and Herpes Simplex Virus Type 2 Infection in Men in the United States: the National Health and Nutrition Examination Survey (NHANES), 1999–2004." *Sexual Transmitted Diseases* 34: 479–84.

Youn, Anthony. 2013. "The Present and Future of Breast Enhancement." *Huffington Post*, March 13. To access: http://www.huffingtonpost.com/anthony-youn-md-facs/the-present-and-future-of_1_b_2864541.html.

Young, Iris Marion. 1992. "Breasted Experience." pp. 215–30 in *The Body in Medical Thought and Practice*, edited by Drew Leder. Boston: Kluwer.

Young, Susan. 2013. "The Latest Artificial Heart: Part Cow, Part Machine." *MIT Technology Review*, May 30.

Zane, Kathleen. 2001. "Reflections on a Yellow Eye: Asian I(\Eye/)Cons and Cosmetic Surgery." In *Talking Visions: Multicultural Feminism in a Transnational Age*, edited by Ella Shohat. Cambridge, MA: The MIT Press.

Zhang, F. and K. B. Pasumarthi. 2008. "Embryonic Stem Cell Transplantation: Promise and Progress in the Treatment of Heart Disease." *BioDrugs* 20(6): 361–74.

Zimmer, Carl. 2008. "Why Darwin Would Have Loved Botox." *Discover*, October 15.

Zoske, Joseph. 1998. "Male Circumcision: A Gender Perspective." *Journal of Men's Studies* 6: 189–208.

INDEX

Note: Page numbers in italic indicate figures.

For Product Safety Concerns and Information please contact our EU representative GPSR@taylorandfrancis.com
Taylor & Francis Verlag GmbH, Kaufingerstraße 24, 80331 München, Germany

www.ingramcontent.com/pod-product-compliance
Ingram Content Group UK Ltd.
Pitfield, Milton Keynes, MK11 3LW, UK
UKHW022217190926
475702UK00003B/398

* 9 7 8 0 4 1 5 8 2 1 9 5 7 *